DEVELOPMENTAL AND CELL BIOLOGY SERIES

ANALYSIS OF LEAF DEVELOPMENT

ANALYSIS OF LEAF DEVELOPMENT

ROMAN MAKSYMOWYCH

Professor of Biology, Villanova University, Pennsylvania

FOREWORD BY

RALPH O. ERICKSON

Professor of Biology, University of Pennsylvania

CAMBRIDGE

AT THE UNIVERSITY PRESS

1973

Published by the Syndics of the Cambridge University Press
Bentley House, 200 Euston Road, London NW1 2DB
American Branch: 32 East 57th Street, New York, N.Y. 10022

Library of Congress Catalog Card Number: 72–83585

ISBN: 0 521 20017 2

Printed in Great Britain
at the University Printing House, Cambridge
(Brooke Crutchley, University Printer)

Contents

Foreword

A number of years ago, it seemed to Maksymowych and me that curiously little was known in a definitive way about the development of leaves, considering their importance as photosynthetic organs, their manifold adaptations to particular climatic and ecological situations, the information which they provide for plant taxonomy, genetics and the study of natural variation, and, for that matter, for one's esthetic appreciation of plants. There were a number of studies such as those of Schüepp and Foster, correlating the form of mature foliar organs (e.g. winter bud scales, photosynthetic leaves) with that of the primordia from which they originated. These studies struck me as having more in common with the ancient idealistic morphology of Goethe, the doctrines of metamorphosis and serial homology, than with the modern objective of developmental biology, to try to find mechanistic explanations of developmental processes. There had also been detailed histological descriptions of shoot apices and the youngest leaf primordia, notably by Esau, which were largely concerned with the question of leaf initiation, or determination, and the origin of the procambium. Developing leaves had also been studied at later stages than that of the youngest primordia as by Avery in *Nicotiana*, and a terminology for the meristems concerned in leaf histogenesis, owing largely to Schüepp, was current.

However, we felt that the available information gave one little feeling for the dynamic nature of the processes involved in leaf development, its growth, if you will. (Richards and Kavanagh's analysis of Avery's data on the growth of a marked leaf of *Nicotiana* should, however, be noted.) This is not to say that the literature relating to leaf form and development was not voluminous, only that we could easily phrase questions, for which answers were not to be found.

Maksymowych has now provided a coherent account of leaf development, based largely on his own research on *Xanthium*. The emphasis throughout is on the dynamics of the growth processes. In his own work, he has used several techniques which are not usual in plant morphological research. The choice of *Xanthium* itself is worth noting. As Ashby had shown, the development of the leaves of many plants is markedly heteroblastic, probably often in response to the photoperiodic regimen. *Xanthium* has a pronounced flowering response to short-day treatment, and the assurance that it will develop stricly vegetatively when maintained on long photo-periods, is certainly convenient in studying vegetative leaf development. In other ways, too, he has been at pains to characterize his plant material more carefully than is usual in plant anatomy. He has made use of controlled plant growth chambers, and using the plastochron index, he has specified the developmental age of his material with statistical precision. These techniques have allowed him to ascertain the time rates of various morphogenetic processes, and to make valid comparisons of these rates, as in his discussion of cell division, cell enlargement, and the expansion in thickness of the maturing lamina. The radioactive labelling techniques which he has used also yield rate data, the rates of incorporation of labelled nucleoside, of course, correlating with rates of cell division.

Using these techniques, he has, for instance, made a more thorough analysis of the roles of differential cell division and expansion, in the maturation of the epidermis, palisade and spongy mesophyll, with their distinctive cell morphologies, than was possible for Avery, using strictly histological techniques. This approach to a study of plant development should be widely applicable, and the value of this monograph should partly be to exemplify methods of dealing effectively with the dynamic aspects of a developmental problem.

It may be that the leaf of *Xanthium* is now better known than that of any other plant, but certainly many questions about leaf development remain to be answered. The simple, entire leaf of *Xanthium* can fairly be regarded as 'typical' in form. However, there is a great diversity of mature leaf forms among the vascular plants. Presumably there is a corresponding diversity in developmental patterns, and there should be some value in comparative studies similar to these of *Xanthium*, to provide a basis for judging which aspects of leaf development are general and which are specialized.

Maksymowych has largely been concerned with questions of morphogenesis at the light microscope and macroscopic levels of organization. Much is being learned today about the ultrastructural aspects of plant morphology, and the role of cell organelles in developmental processes. It should be clear from this book that the leaf presents several interesting questions for electron microscopic study, and I would think that precise specification of the developmental status of the material studied, would be as important in ultrastructural studies, as it has been in Maksymowych's work.

Leaves have many biosynthetic capacities, of which photosynthesis is, of course, the most important, but the synthesis of sugars, fatty acids, amino acids, proteins, polysaccharides, alkaloids, and plant growth substances come to mind. Very little is known about the development of these synthetic capacities. In the case of photosynthesis, much work has been done on the differentiation of chloroplasts in etiolated leaves which are returned to light, and the concomitant recovery of photosynthetic activity by these leaves. It is not clear to me how this course of events relates to the development of the proplastids of a leaf primordium into the functional chloroplasts of the mature leaf, and to the acquisition of photosynthetic capacity by the leaf, during its normal development. The information provided in this volume should form a sound basis for studies of the epigenesis of the biosynthetic mechanisms of a leaf.

More abstruse questions concerned with induction, determination and the control of developmental processes of the leaf, and other plant organs, are almost completely unanswered, and are therefore difficult even to discuss. We do have some general understanding of the roles of several growth substances, such as auxin, the gibberellins and phytochrome. Certainly they have profound morphogenetic effects, but there has scarcely been any attempt to explain these effects in terms of the effect of the growth substances on meristematic function, and on patterns of differentiation, which the morphologist sees as important. Advances in this area of plant physiology can certainly be anticipated. However, in my opinion, satisfactory explanations of developmental processes will consist, not of pointing out simple correlations of cause and effect, but in finding mathematical relationships, which express the interplay of endogenous and environmental factors, with the form of an organ. Feedback relationships, as in the servo-mechanisms and control circuits of engineering, will perhaps be an important component of future theories. If this prediction has any validity, I should think that the description of leaf development, which Maksymowych gives here, with its emphasis on the dynamics of the processes involved, should be a good basis for future study.

Department of Biology
University of Pennsylvania

Ralph O. Erickson

Preface

Since the appearance of Avery's classical paper on tobacco leaf development in 1933, much developmental work has been done with tissue culture and various plant hormones. Relatively little research has been carried out with intact leaves following an undisturbed course of development. This imbalance is evident in the present day textbooks. To name a few examples, little data is available on quantitative description of leaf growth, DNA synthesis, cell division, respiration and chlorophyll synthesis at various stages of leaf development.

A comprehensive up-to-date review of leaf development would appear to be a worthwhile contribution. This monograph should specifically fill the existing gap. It is experimental and quantitative; it contains morphological, cellular and metabolic aspects of development and it covers the entire course of development. Graphical representation of processes and the use of plastochron index as a developmental scale provide a good basis for correlation of various processes.

This monograph comprises more than ten years of research which has been published in various journals but now is brought together under one unified theme. Most of the work was done with *Xanthium* leaves but a significant proportion of developmental studies on other genera is also included.

R.M.

Acknowledgments

I would like to express my gratitude to Professor R. O. Erickson who over a number of years has stimulated my interest in plant development. The quantitative approach adopted in this book owes a great deal to his influence. His ready advice and constructive criticism of the manuscript is gratefully acknowledged.

I am also indebted to Professor John G. Torrey of Harvard University for his interest and constructive criticism of the manuscript.

Rev. R. C. Shurer, O.S.A. of the Biology Department at Villanova University has given his generous help and advice in microphotography. For this I am grateful.

Villanova 1972 R.M.

I

MORPHOLOGICAL PATTERNS OF DEVELOPMENT

1

Plastochron index

A plastochron index was proposed for the study of vegetative shoot development of *Xanthium italicum* Moretti (cocklebur) by Erickson and Michelini in 1957. This index and a similar leaf plastochron index are linearly related to time, and consequently they provide an indirect time scale which may be used in developmental studies. Since the leaf plastochron index was used in the quantitative description of leaf development it may be appropriate to explain briefly its derivation and the assumptions upon which it was derived.

Xanthium plants were grown under greenhouse conditions and received 16 hours of light per day to maintain the vegetative state. They are typical short day plants, and will flower if exposed to even one dark period which exceeds 9.5 hours of the 24 hour cycle. The total length, including lamina and petiole, of each leaf was measured daily from July 8 to August 26. A family of sigmoid curves was obtained and when leaf length was plotted logarithmically against the time as in Fig. 1, the relationship between logarithm of leaf length and time was linear up to at least 70 mm leaf length. This would indicate that leaves are growing exponentially in their early stages of development. It is also evident from data presented in Fig. 1 that leaves are initiated at approximately equally spaced time intervals, judging from spaces between lines, and that the lines were roughly parallel to each other.

A plastochron can be defined as the time interval between initiation of any two successive leaves. More broadly it can also be defined as the interval between corresponding stages of development of successive leaves at their initiation, maturity, or any other stage of development which can be used as a reference stage. A length of 10 mm is convenient for *Xanthium* since at this length leaves are growing exponentially and they are large enough to be measured with some accuracy. This reference length is designated with a dashed horizontal line in Fig. 1. A plastochron may be visualized as the distance between the intersections of the reference line with the growth curves of any two successive leaves. Hence a cocklebur plant which has five intervals on the dashed line will be five plastochrons old. Specifically a plant will be n plastochrons old when leaf n is 10 mm long. When, however, no leaf is exactly 10 mm long an interpolation is required for an exact estimation of the plastochron age of the shoot. Exponential growth in the early stages of leaf development, equally spaced initiation of leaves and

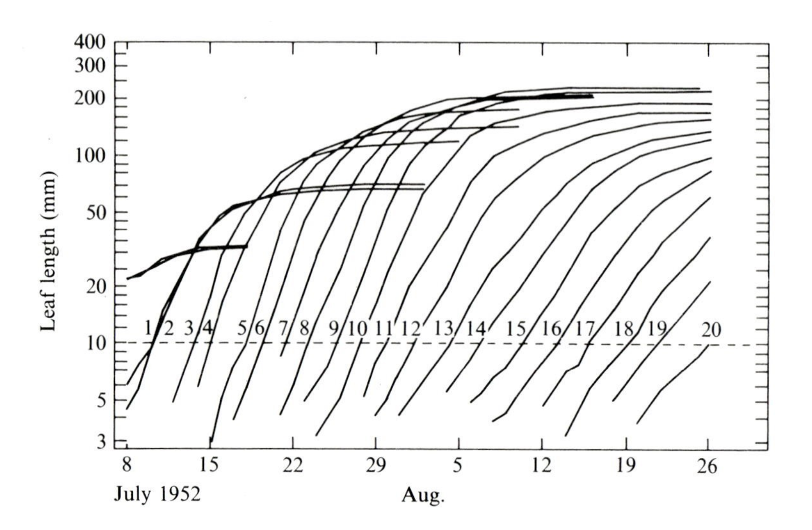

Fig. 1. Lengths of successive leaves of a *Xanthium* plant plotted logarithmically against time. Each growth curve applies to a single leaf of the same plant. The unmarked curves at the left apply to the two cotyledones. (From Erickson and Michelini, *Amer. Jour. Bot.* **44**, 1957.)

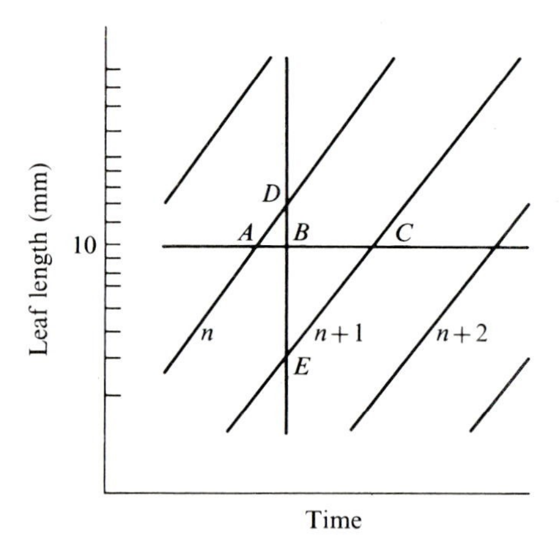

Fig. 2. Idealized geometric representation of a portion of Fig. 1. (From Erickson and Michelini, *Amer. Jour. Bot.* **44**, 1957.)

parallel profiles of the family of lines were the basic assumptions upon which interpolation could be made and the fractional values of the plastochron determined.

Fig. 2 is a diagrammatic representation of a portion of Fig. 1 in which the diagonal lines represent parts of the growth curves and the horizontal line is the reference line drawn at 10 mm leaf length. A vertical line is drawn at a time when the leaf n is longer than 10 mm and leaf $n+1$ is shorter than the established reference length of 10 mm. In this case, a fractional part of the plastochron must be established. Triangles ABD and CBE are similar and $AB/AC = DB/DE$. The interpolated value of the fractional plastochron is $n+(AB/AC)$. However, the values DB/DE are more convenient to use since they represent leaf length before and after the time specified by the vertical line. The fractional plastochron is now $n+(DB/DE)$. Since logarithms of leaf lengths are used, DB is equal to $\log L_n - \log 10$, and DE is equal to $\log L_n - \log L_{n+1}$. An equation for the plastochron age of the plant can now be formulated.

$$PI = n + \frac{\log L_n - \log 10}{\log L_n - \log L_{n+1}}.$$

PI represents plastochron index which is the age of the plant expressed in plastochrons.

n is the serial number, counting from the base, of that leaf which is just longer than 10 mm.

$\log L_n$ is the logarithm of length (in millimeters) of leaf n, which is longer than 10 mm.

$\log L_{n+1}$ is the logarithm of length of leaf $n+1$ which is just shorter than 10 mm.

To calculate the PI of a plant, it is necessary to count the number of leaves on the stem which are longer than 10 mm and to measure the lengths of the two successive leaves, n and $n+1$, which are respectively, just longer and just shorter than 10 mm. The obtained values are then substituted into the PI equation.

The leaf plastochron index (LPI) can be used in developmental studies limited specifically to only one leaf. To estimate LPI, the serial number of the leaf in question is subtracted from PI of the plant which bears it, $LPI = PI - a$, where a is the serial number of the desired leaf. Any leaf longer than 10 mm will be of positive LPI, and any leaf shorter than 10 mm will be of negative plastochron age. A leaf which is exactly 10 mm long will be of zero LPI.

It may be convenient at this point to illustrate one example of estimation of both PI and LPI.

From measurements on one cocklebur plant the following data were obtained:

$$
\begin{aligned}
n &= 11,\\
L_n &= 16 \text{ mm},\\
L_{n+1} &= 8.8 \text{ mm},\\
\text{PI} &= n + \frac{\log L_n - \log 10}{\log L_n - \log L_{n+1}}\\
&= 11 + \frac{1.2041 - 1.0000}{1.2041 - 0.9442}\\
&= 11.79, \text{ which represents the plastochron age of the entire plant.}
\end{aligned}
$$

Suppose we are studying the development of leaf 9 of the same plant and we would like to estimate the age of that leaf in plastochron units (LPI).

$$
\begin{aligned}
\text{LPI}_9 &= \text{PI} - 9 = 11.79 - 9\\
&= 2.79
\end{aligned}
$$

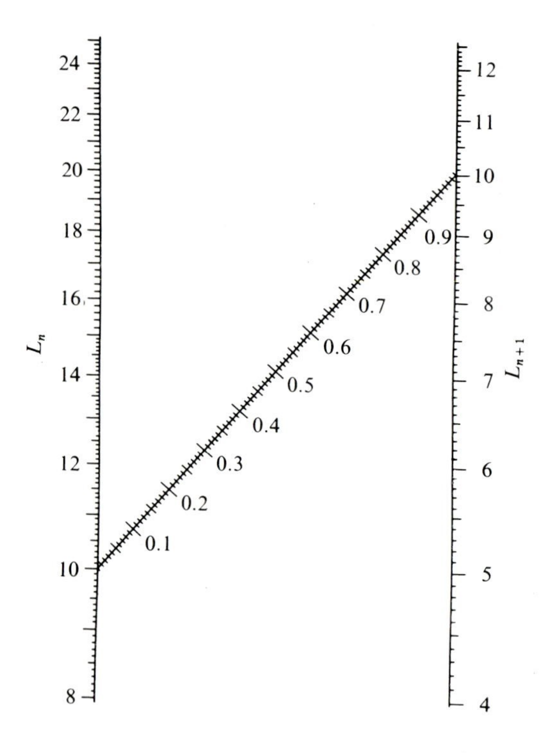

Fig. 3. Nomogram which can be used for quick estimation of plastochron ages of a shoot and leaves. (From Erickson, *Amer. Jour. Bot.* **47**, 1960.)

To simplify estimation of the fractional value of this equation Erickson (1960) suggested the use of a nomogram which is illustrated in Fig. 3. The nomogram consists of two parallel logarithmic scales, L_n and L_{n+1} with a diagonal scale connecting the reference leaf lengths of 10 mm. If a ruler is placed on 16 and 8.8 values of the vertical scales, the fractional plastochron value estimated from the diagonal scale is 0.783. This agrees fairly well with the number derived from the equation.

It seems an appropriate question to ask whether it is really necessary to use the plastochron index in developmental studies, especially when controlled environmental conditions are used. The practicality of the use of this index was discussed by Michelini (1958). He demonstrated that the use of the plastochron index offers many advantages in interpretation of physiological processes during plant growth and development, over the use of chronological time in similar studies. It can be pointed out that leaf plastochron index should be considered as a precise developmental scale, perhaps similar to the abcissa of a co-ordinate system. This scale centers at zero, when the leaf is 10 mm long. It acquires negative values at younger stages and a positive value when the leaf grows older. With this scale it is possible to establish a base line in time which can be used to estimate in plastochron units when a leaf primordium is initiated, when the lamina is initiated, when DNA synthesis stops and perhaps to establish the period of maximum rate of cell enlargement. There is also an advantage in the timing of development and selection of desired developmental stages, especially when leaves have to be sacrificed for experimental use. Theoretically, one could even estimate the age of a leaf primordium even before any visible evidence of its initiation.

2

Organization of the shoot apex

The terminal shoot of *Xanthium pennsylvanicum* (Fig. 4) is composed of the stem apex with its apical meristem, leaf primordia, young leaves, axillary buds and stem. This description obviously is based upon the external morphology. Functionally, the meristems of the stem apex produce leaf primordia which develop into mature photosynthetic organs and the stem proper. This happens under conditions suitable for vegetative growth. After photoperiodic induction, the vegetative growth is curtailed and the stem apex is transformed into a flower bud with subsequent sexual reproduction (Salisbury, 1961). However, our attention will be focused on the apical organization during the vegetative reproduction.

According to Millington and Fisk (1956), Wetmore, Gifford and Green (1959) and Gifford (1963), the *Xanthium* stem apex appears to be composed of the one-layered tunica with large prominent cells, and the corpus beneath it, characterized by smaller cells. However, Schmidt's tunica–corpus concept should be applied here only loosely. Some stem apices, as illustrated in Figs. 4 and 5, have two distinct layers of large cells which could be classified as tunica but cells of the second layer may divide periclinally. Gifford (1963) made similar observations concerning the two conspicuous layers of large cells. However, according to his description the second layer is a part of the corpus, perhaps because of periclinal divisions.

The region of dividing cells at the tip of the shoot constitutes the apical meristem. Topographically, this meristem can further be divided into a central zone, peripheral zone and rib meristem zone. The central zone is located at the summit of the stem apex and is characterized by large vacuolated cells with least frequent cell division. In the peripheral zone, which is located at the flanks of the apex, cells are smaller with dense cytoplasm; they divide more frequently at almost any orientation of the cell plate. This meristem is directly responsible for initiation of leaf primordia. Autoradiographic studies (Bernier, 1966) of apical meristems labeled with ^{3}H-thymidine and ^{3}H-adenine revealed that DNA and RNA synthesis is more active in the peripheral than in the central zone and in the rib meristem. The density of ribosomes per unit area is also higher in the cytoplasm of the peripheral cells if compared to the other zones of the apical meristem. The production of leaf primordia therefore appears to be well correlated with higher syn-

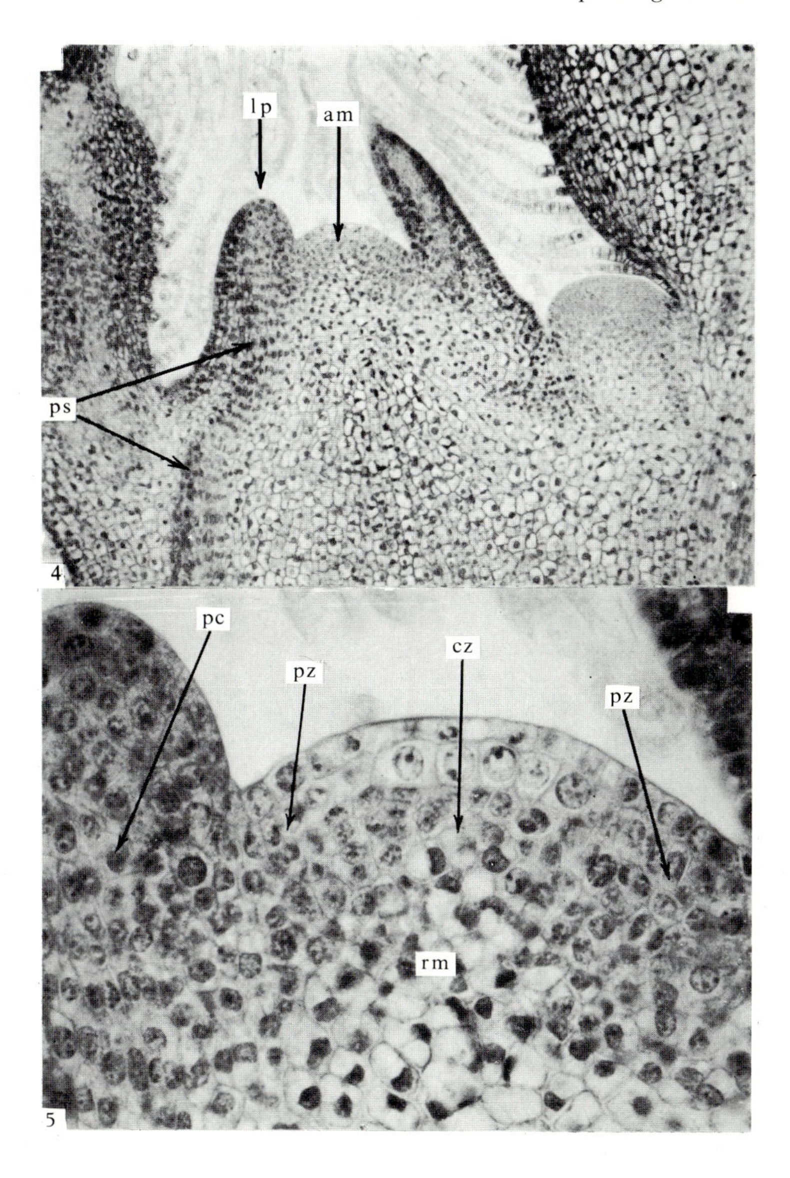

Fig. 4. Shoot apex of *Xanthium pennsylvanicum* sectioned median to the youngest leaf primordium (100×); lp indicates leaf primordium, ps – procambial strand and am – apical meristem.

Fig. 5. Enlarged portion of the apical meristem and the youngest leaf primordium (430×) showing procambial cells (pc), peripheral (pz), central (cz) and rib meristem zone (rm).

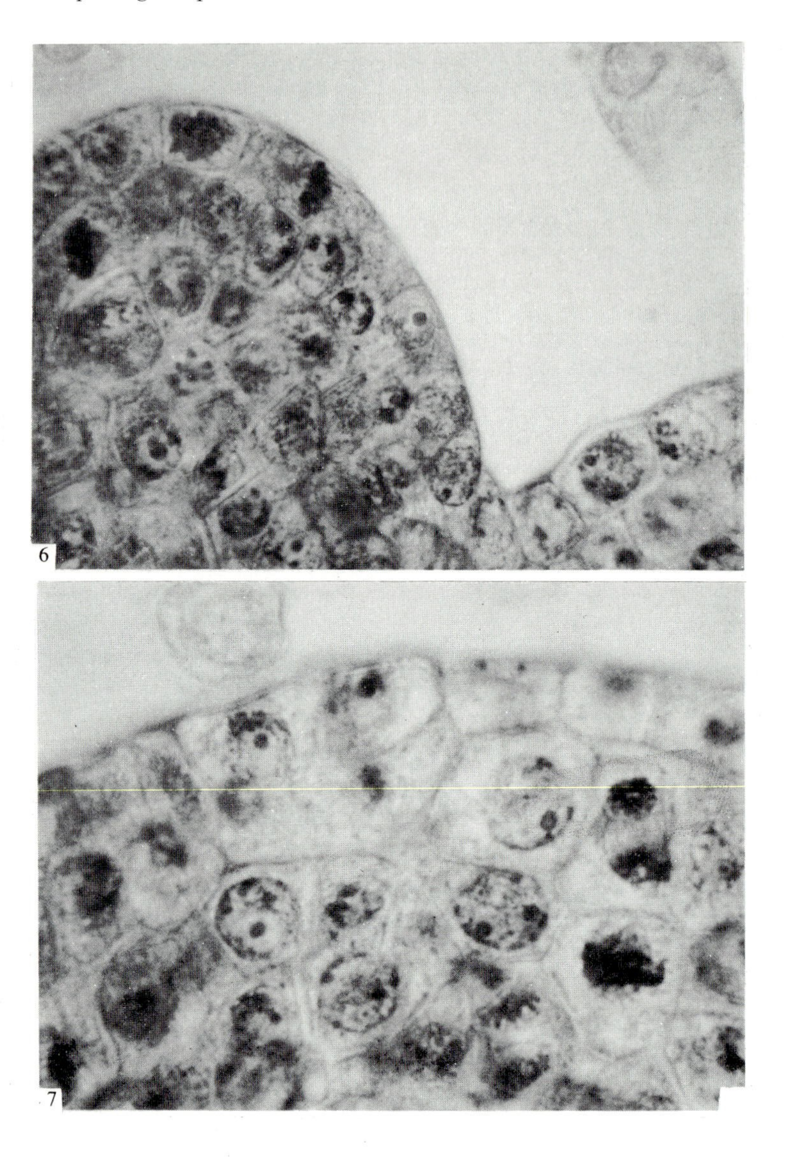

Fig. 6. Enlarged youngest leaf primordium with mitotic division in its apical meristem (900 ×).

Fig. 7. Portion of the apical meristem magnified 1350 × with periclinal cell divisions in the second and third layers of the stem apex, indicating a possible initiation of a new leaf primordium.

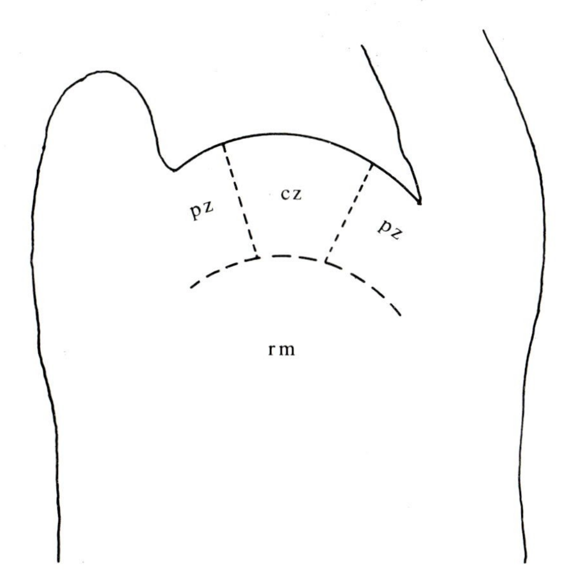

Fig. 8. Diagram of the apical meristem marked to show the peripheral zones (pz), central zone (cz) and rib meristem zone (rm).

thetic activity of the peripheral zone. The pith rib meristem is contiguous with the basal part of the central zone. A preponderance of periclinal divisions results in the formation of files of cells typical of the rib meristem. A schematic representation of the zonation pattern is illustrated in Fig. 8. Buvat (1952, 1955) and Lance (1952) have advanced a concept that the center of the apical meristem is '*peu active*' and only the peripheral zone is highly active in vegetative growth. In functional zonation of the shoot apical meristem based upon mitotic index studies, they distinguish the '*méristème d'attente*' – quiescent or latent meristem – which would be equivalent to the central zone; the '*anneau initial*' to the peripheral zone; and the '*méristème medullaire*' equivalent to our rib meristem. The *méristème d'attente* is considered to be mitotically inactive, but the latter two meristems, according to this concept, are extremely active during vegetative growth and are self perpetuating. Based upon cytochemical studies (Lance, 1952; Gifford, 1963) and mitotic indices, the *méristème d'attente* cannot easily be recognized in *Xanthium*. Millington and Fisk (1956) estimated the relative number of mitotic figures from median longitudinal sections of 20 *Xanthium* shoots. They found 3.2% of mitosis in the tunica layer, 0.8% in the central zone, 16.1% in the peripheral zone and 4.9% in the rib meristem. The data indicate that while cell division is most frequent in the peripheral zone, the central zone participates actively in the development of the shoot apex.

Millington and Fisk (1956) also made some interesting observations on the shoot apex of the mature embryo in the seed. Such an apex is somewhat smaller and more flattened than that of the germinated plant. The cells of the embryo are more uniform in size than those of a developed shoot. Since the cell size of both meristems is similar, the lesser volume of the embryonic shoot can be attributed primarily to fewer cells. The marked zonation of the apical meristem is not apparent in the embryo, but becomes distinguishable during the early ontogeny of the seedling. A comprehensive discussion on the zonation patterns of stem apices is presented by Clowes (1961) in his book on apical meristems.

3

Leaf initiation

As stated before, one of the functions of the apical meristem is to produce leaves. The peripheral zone of the meristem is actively engaged in this process. Cellular initiation of leaf primordia in *Xanthium* follows a pattern similar to that of many other genera (Cutter, 1964; Esau, 1965; Kaplan, 1970). Periclinal cell divisions in the second and third layers of the stem apex are earliest indications of leaf inception. Early mitotic telophase and metaphase are illustrated in Fig. 7. A center of mitotic activity is established at a precise phyllotactic location on the apex with a precise temporal relationship to other initiated leaves. The phyllotactic leaf arrangement in *Xanthium* can be represented with a 2,3 contact parastichy pattern. Leaf primordia are initiated at approximately 3.5-day intervals. Cells of the epidermal layer divide anticlinally and together with the underlying tissue they establish a small protuberance, frequently referred to as the leaf buttress (Esau, 1965). The newly initiated leaf primordium grows by cell division and cell enlargement of its apical meristem, which is established at the tip (Figs. 6 and 10), and also by intercalary growth, again consisting of cell division and cell enlargement. The youngest primordia can be recognized easily on a transversely sectioned apex at LPI -6.0 (Fig. 21). Leaf initiation, therefore, occurs before LPI -6.0, and can be recognized in longitudinal sections in Fig. 7. As the young primordium grows in height, it becomes somewhat flattened and the central part of it develops into the midrib of the future leaf. As soon as the primordium has reached a height of about 60 μm formation of procambium is noticeable. It appears that the procambial strands (Figs. 4 and 5), established in the stem at the periphery of the pith rib meristem, differentiate acropetally into the newly formed primordium. Presence of procambium together with periclinal cell divisions was interpreted by McGahan (1955) as a signal of leaf initiation. During the subsequent development, the established procambial strands differentiate into xylem on the adaxial side and phloem on the abaxial side of the leaf, respectively. McGahan (1955), describing vascular differentiation in *Xanthium chinense*, concluded that maturation of xylem elements in a young leaf occurs acropetally, but the contact with the mature xylem in the stem is established by basipetal differentiation of the procambial strands. The pattern of vascular differentiation in *Xanthium* appears to be similar to that in *Helianthus* and

Sambucus (Esau, 1965). Jacobs and Morrow (1957) studied xylem development in the vegetative shoot of *Coleus*. Their findings were similar to those of earlier investigations. Acropetal phloem differentiation preceded the development of xylem tissue. The first pair of leaf primordia had no mature xylem elements. Discontinuous xylem strands were found infrequently in the second pair, but most frequently in the third pair of leaf primordia. The first locus of xylem differentiation was near the base of young primordia. Differentiation occurred both in the acropetal and basipetal direction. All leaves longer than 1600 μm had xylem strands extending continuously from their distal site down through the leaf and into the stem. The authors found a close correlation between leaf length and various stages of xylem differentiation. A similar correlation was found between the auxin level and xylem differentiation in the vegetative shoot of *Coleus*.

4

Origin and development of the lamina

The youngest primordium recognizable in a given transverse section shows no evidence of lamina initiation (Fig. 9, LPI −5.11). In 24 out of 26 apices, (Maksymowych and Erickson, 1960) the second youngest primordia have developed laminae to some extent or at least show some indication of its initiation. The two squares in Fig. 21 around LPI −5.0 show the first indication of marginal activity; the other two squares reveal no lamina initiation, which can be easily detected. Leaf primordia older than LPI −5.0 show more advanced stages of leaf blade development. It can be stated therefore that lamina initiation in *Xanthium* occurs at LPI −4.8. At this stage the primordium is about 220 μm long. The lamina is initiated from multiple rows of meristematic cells located along two margins of the leaf axis (Fig. 9; LPIs −5.0 to −3.6). These rows of cells are called the marginal meristem. For *Xanthium*, the marginal meristem can be defined as a group of cells within four cell diameters of the leaf margin which initiates the leaf blade by forming five or six basic cell layers of the future plate meristem. Marginal meristem is characterized by a preponderance of anticlinal and oblique cell divisions, showing a pattern distinct from the plate meristem. As assayed by the mitotic counts (Fig. 13) and ^{3}H-thymidine incorporation into nuclear DNA, this meristem is active from LPI −4.8 to LPI +2.0 for about 23 days.

The plate meristem of the *Xanthium* leaf usually consists of five or six basic cell layers in the early stages of development and is characterized by a preponderance of anticlinal cell divisions, which bring about the extension of the leaf blade. This meristem extends from the midrib to the leaf margin, however the continuity of the cell layers is interrupted by formation of procambial strands and veins in subsequent development. At some later stage of development, periclinal cell divisions increase the number of cell layers, perhaps within the range from 5 to 10, and some growth in thickness takes place. The periclinal and oblique divisions of the plate meristem undoubtedly also gives rise to procambial strands. The total duration of mitotic activity is about 23 days, from LPI −4.0 to +2.8, after initiation of the six basic cell layers. As illustrated in Fig. 13, the pattern and the rates of cell division are significantly different from those in the marginal meristem. The concept of the marginal meristem has not yet been formulated with sufficient clarity for other plants than *Xanthium*. Perhaps because of the

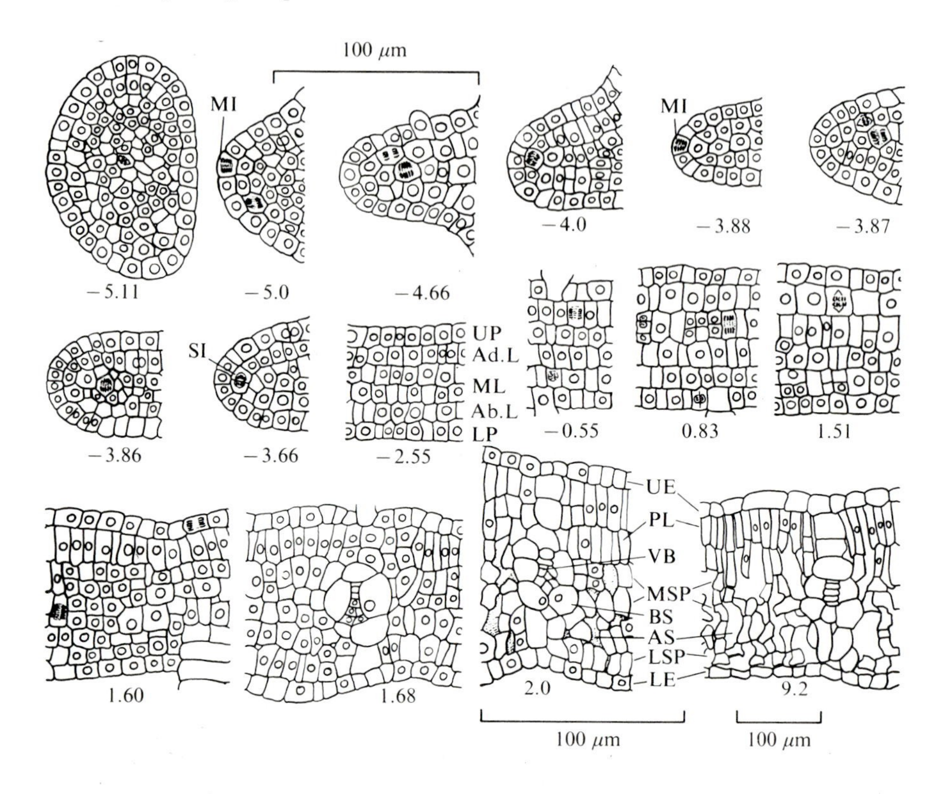

Fig. 9. Transverse sections of a young leaf primordium, of the marginal meristem of older primordia, and of portions of the lamina of still older primordia and leaves. The number below each drawing is the LPI of the leaf from which the section was obtained. (MI – marginal initial region, SI – submarginal initial region; UP and LP upper and lower protoderm, Ad.L, ML and Ab.L – adaxial, middle and abaxial layers of the immature lamina; PL – palisade parenchyma; VB – vascular bundle; BS – vascular bundle sheath; MSP – mesophyll parenchyma; AS – intercellular space; LSP – lower spongy parenchyma; UE and LE – upper and lower epidermis).

variability in patterns among different genera, no one has offered a general discussion of spatial and cellular delineation of the extension of this meristem from the leaf margin. Further, no definite answer is provided as to how long during development it remains active. Some workers (Avery, 1933; Denne 1966; Fuchs, 1966) have stated that active marginal growth is of short duration, but this is not supported by data for the whole period of development and is contrary to present observations in *Xanthium*.

It is generally held that the plate meristem shows chiefly anticlinal divisions and that the number of cell layers originally established in the young organ does not increase; thus an extensive sheet of tissue is produced

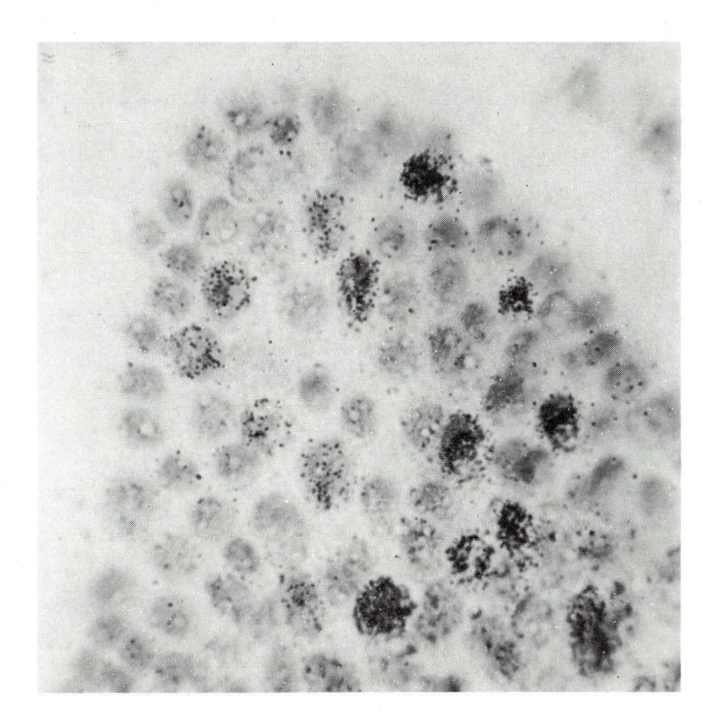

Fig. 10. Autoradiograph of a young leaf primordium showing cells labelled with tritiated thymidine. The grains over some nuclei represent relative amounts of ^{3}H-thymidine incorporated into DNA. Both, apical and intercalary growth are contributing to the extension of the primordium.

(Esau, 1965; Schüepp, 1966). The growth of the lamina in thickness is another aspect of cell division in *Xanthium* leaves which is not consistent with the above description (Maksymowych and Wochok, 1969). As a result of continuing activity of the plate meristem, 7 to 10 layers are established by numerous periclinal divisions at about LPI 1.5 from the five or six cell layers. This occurs before rapid differentiation of the laminar tissue but may also involve some enlargement of cells. An example of this developmental stage is illustrated in Fig. 17. It appears that in addition to the surface growth of the lamina by anticlinal divisions, the plate meristem contributes also to the growth in thickness by numerous periclinal divisions. There is no reason to define the stage of periclinal and oblique divisions as constituting another meristem, distinct from the plate meristem.

In addition to *Xanthium* data (Maksymowych and Erickson, 1960), Denne (1966) has provided supporting evidence from her work on development of *Trifolium* leaves. About 30 days after the establishment of four cell layers by anticlincal divisions of the plate meristem, some cells divide periclinally, to produce a two-layered palisade. The fourth layer of spongy mesophyll is also formed by periclinal divisions in one of the two innermost layers of the lamina. Thus the plate meristem contributes more cell layers and some growth in thickness takes place.

Present day concepts of marginal and plate meristems would have to be extended and perhaps partially modified to describe the activity of these meristems in *Xanthium* and *Trifolium* leaves.

5

Cell lineage in the marginal and plate meristems*

In studies of the histogenesis of the lamina of other species, several authors (Weidt, 1935; Foster 1936; Gifford, 1951; Girolami, 1954) have summarized observations on lamina histogenesis in the form of cell lineage diagrams. In their reports arrows were used to indicate the derivation of the upper and lower protoderm from the marginal initials; of the adaxial, middle and abaxial mesophyll layers from the submarginal initials; and the mature tissues from these meristematic layers. Noack (1922) made a series of hypothetical drawings of the transverse section of the leaf margin in *Pelargonium*, purporting to indicate the cell lineage from the initials. Avery (1933) and several other authors have made statements about cell lineage without using diagrams.

From the studies of the sections of *Xanthium* leaves, Maksymowych and Erickson (1960) have failed to understand the basis on which one can draw conclusions about cell lineage with the certainty which such diagrams imply. Where it is possible to keep individual growing cells under observation for a period of time (Goodwin and Avers, 1956) the derived cell lineage would be unequivocally clear. In dealing solely with parallel specimens which have been fixed, one can relate the cells of one specimen with those of another only on the general grounds of position and form, and only indirect evidence can be obtained about cell lineages. Only limited conclusions can be drawn from such indirect evidence. Perhaps the most objective, though indirect, evidence which such sections provide is in the orientation of the new cell plates formed in cell division. While wall orientation in general may be indicative of growth processes occurring in meristems (Hejnowicz, 1955) it may be advisable to limit observations to newly formed cell plates, to avoid the uncertainties which later reorientation of older cells might introduce. Maksymowych and Erickson (1960) measured the angular position of the cell plate in each metaphase, anaphase and telophase figure seen in the marginal position of a number of transverse sections of the lamina of several leaves of known plastochron age. In the case of the surface layer of cells, like protoderm and epidermis, the position of the cell plate was specified by

*Part of the discussion of the cell lineage in the marginal and plate meristems was contributed by Prof. R. O. Erickson. This contribution is gratefully acknowledged.

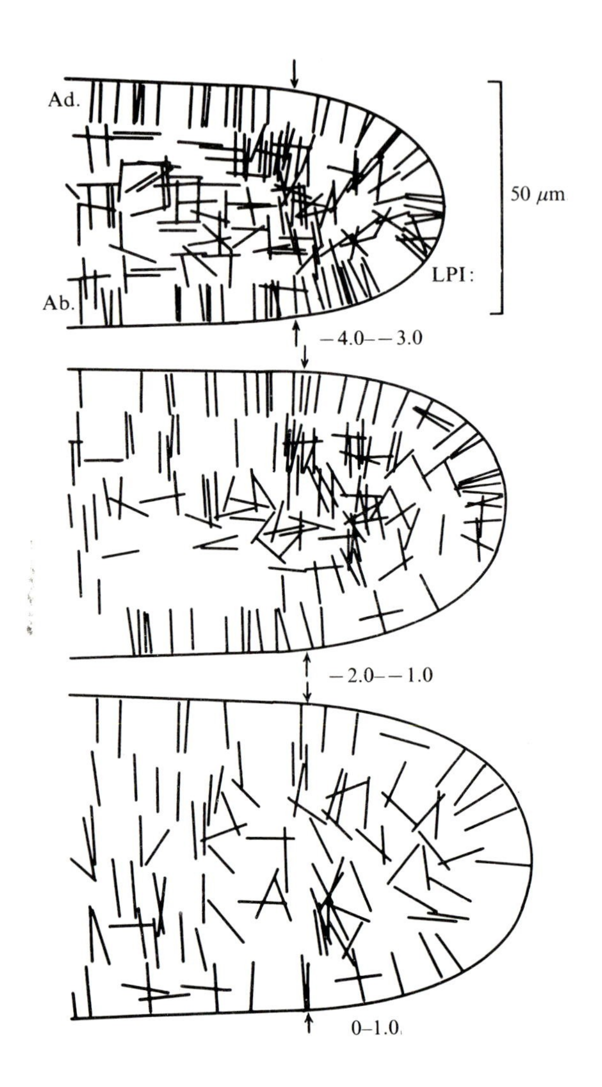

Fig. 11. Composite diagrams showing orientation of cell plates in numerous transverse sections of leaf margins, at three LPI intervals. Adaxial surface (Ad) is at the top of each diagram. Small vertical arrows delineate marginal meristem from plate meristem.

the angle which it made with a line drawn at a tangent to the surface wall of the mother cell. Mere inspection of Fig. 11 shows that the great majority of cell plates in the protoderm are either approximately perpendicular, or approximately parallel to this surface. For the purpose of statistical analysis, the range of the angular position of the cell plates was divided into three classes: those which were perpendicular to the specified surface ($\perp$; 67.5 to 112.5°); those which were oblique ($<$; 22.5 to 67.5° and -67.5 to -22.5°); and those which were parallel ($\|$; -22.5 to 22.5°). These classes may correspond to conventionally defined anticlinal, oblique and periclinal types of division, respectively. In all three drawings of Fig. 11, 141 cell plates were recorded in the protoderm. Of these, 128 were classified as perpendicular to the surface or anticlinal, and 11 as parallel or periclinal, giving proportions 0.91 and 0.08, respectively. In a statistical sense, these proportions represent a highly significant departure from a random arrangement. In the random arrangement of the cell plates the expected proportions would be 0.25 for the perpendicular and parallel classes, and 0.50 for the oblique class. These data clearly imply that the protoderm with its preponderance of anticlinal cell division is a histogen in the classical sense. The cells of this histogen are all derived from initial cells at the margin and they do not contribute new cells to deeper tissues of the leaf. The periclinally oriented cell plates appear to be involved exclusively in the formation of epidermal hairs.

The proportion of perpendicular, oblique and parallel cell plates of the subepidermal cells nearer the margin was found to be significantly different from the proportion of cell plates situated farther from the margin. These differences appeared most pronounced when a dividing line was placed at about 4 cell diameters from the margin. This dividing line is indicated by arrows in Figs. 11 and 12. It is approximately the region where six basic layers of the plate meristem are about to be established. In Fig. 14, the proportions of perpendicular, oblique and parallel cell plates in these two regions are indicated by histogram bars. These proportions may be compared with those to be expected if cell plates were arranged at random. In the more marginal group of cells, 111 in all, the proportion of perpendicular cell plates, 0.42, is significantly higher than 0.25 which would be expected on the random hypothesis. The proportion of parallel cell plates, 0.153, is significantly lower than 0.25, whereas the proportion of oblique cell plates, 0.423, cannot be considered to differ from the random figure of 0.50. Of the 171 cell plates found in the subepidermal cells more than about four cell diameters from the margin, there is a great preponderance, 0.58, of cell plates perpendicular to the horizontal plane. However, a very small proportion, 0.116, is oblique to this plane, while the proportion of parallel cell plates, 0.298 is essentially a random proportion. It is clear that in both regions cell plates are preferentially oriented in particular directions and that the pattern of orientation differs between the two regions.

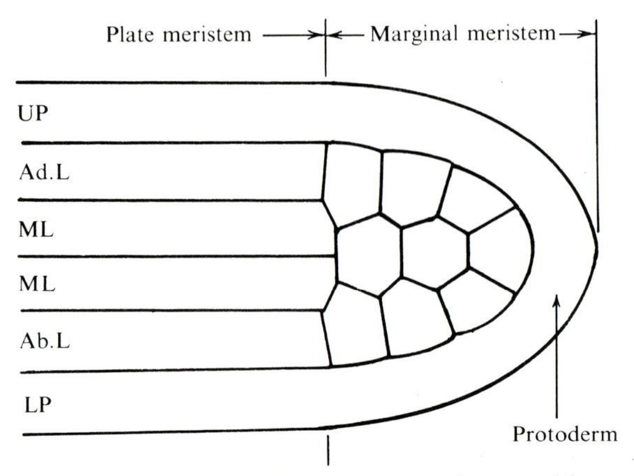

Fig. 12. Diagram representing marginal and plate meristems with upper and lower protoderm (UP, LP), adaxial and abaxial layers (Ad.L, Ab.L) and two middle layers (ML).

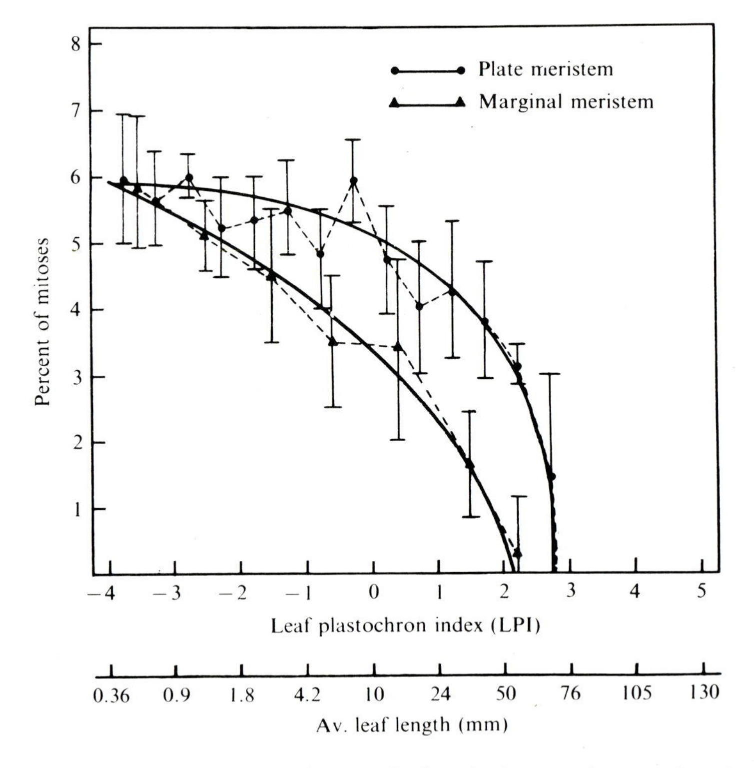

Fig. 13. Percent of mitoses in the marginal and plate meristems plotted vs. LPI. Plastochron ages of leaves on the abscissa are correlated with their respective leaf lengths.

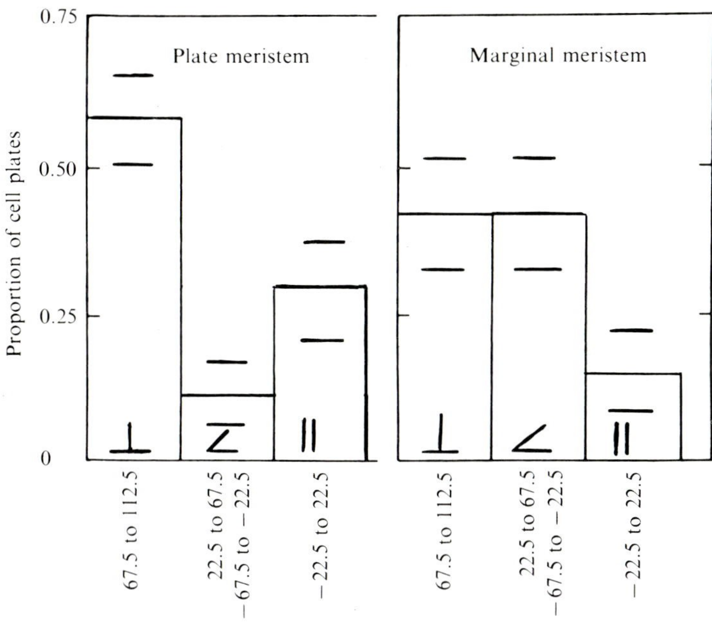

Fig. 14. Proportion of cell plates oriented approximately perpendicular, ⊥; oblique, <; and parallel, ‖ to horizontal plane of leaves. The pattern of orientation of cell plates differs between the marginal and plate meristems.

The region within about four cell diameters of the margin delineated with arrows in Figs. 11 and 12 may be considered as part of the marginal meristem, while the cells more than four cell diameters from the margin might be included in the plate meristem. The plate meristem extends farther inward from the margin than is shown in Figs. 11 and 12. A high proportion of perpendicular cell plates in the two meristems, can be correlated with the lateral expansion of the lamina. Consequently the marginal meristem should probably also be characterized by a high proportion of oblique, and a low proportion of parallel divisions. The plate meristem is characterized by its low proportion of oblique and a high proportion of perpendicular divisions.

Divergence of cell plates in the position of submarginal initial and other individual cells indicates that the pattern of cell division in these regions is somewhat irregular and variable. It is difficult to derive, from data of this sort, unequivocal cell lineages stemming from the submarginal initial cell.

In the region beyond four cells from the margin, where the six basic layers have been established, the proportions of the perpendicular cell plates are significantly higher than the proportions of the expected random arrangement. The anticlinal divisions are certainly associated with the lateral expansion of the lamina, and this region may therefore be considered a part of the plate meristem. Most of the oblique cell plates could be associated with the formation of vascular tissue together with some of the perpendicular and parallel cell plates. The parallel cell plates are probably associated with the increase in the number of cell layers of the mesophyll.

It is perhaps implicit in the concept of the marginal meristem, as well as of other meristems, that its characteristic pattern of organization is maintained for a considerable period of time, and that it can be interpreted as the pattern of developmental processes. This implicit assumption is derived from the analysis of the mitotic planes in the marginal meristem in which there appears to be no significant changes in the orientation of division figures from LPI -4.0 to $+1.0$, or at least 18 days.

Foster (1936), in his review article, distinguished two patterns of marginal activity on the basis of the planes of division of the submarginal initial cells. In one type, illustrated by *Bougainvillea, Pelargonium* and *Nicotiana* 'anticlinal and periclinal division planes alternate in the submarginal initial', and in a second type, illustrated by *Carya* and *Heterotrichum* 'the submarginal initials, by alternating oblique divisions, first produce two internal layers of cells. At varying distances from the leaf margin, the adaxial layer divides periclinally, thus producing an inner or middle tier of cells.' Based on *Xanthium* data, where several hundred sections have been examined, only two division figures have been found in the submarginal position. Since other authors present no data, one fails to see the evidence for the alternating patterns of division. There is no adequate evidence for the explicit and regular cell lineage patterns which various authors have claimed to occur in the marginal meristem. One has to be in sympathy with conclusion of Shushan and Johnson (1955) who say of the submarginal initials of *Dianthus caryophyllus* that 'divisions are perpendicular, anticlinal and oblique but could not be demonstrated to occur in any regular sequence'.

6

Thickness of the lamina

The increase in thickness of the lamina can be attributed to a number of processes which occur in its component tissues. The addition of new layers by cell division and cell enlargement will, of course, be reflected in the thickness of the lamina. It is, therefore, interesting to know the course of expansion of the lamina in thickness, and to correlate it with other developmental features.

Measurements of the lamina thickness were made of fresh and fixed sections. The amount of shrinkage in the processing of the fixed sections in Navashin's solution increased with the plastochron age, reaching a value of about 5% for very young leaves and about 25% for mature leaves. In the early stages of lamina formation at LPI −4.25, as can be seen in Fig. 15, the lamina is about 52 μm thick. At LPI −3.0, when the six layers are clearly established, the thickness is about 60 μm. Then there is a slight progressive increase in thickness up to LPI 1.0 which occurs without any essential increase in the number of cell layers. A very rapid four-fold increase occurs between LPI 1.0 and 4.0, involving both cell division and cell enlargement. The number of cell layers increases rather quickly from the six basic layers up to about eight or nine, or even higher. This sudden growth in thickness takes place between LPI 1.0 and 1.7. Thereafter, the cells of the palisade mesophyll elongate rapidly perpendicular to the plane of the leaf, while the cells of the spongy mesophyll also enlarge, and the intercellular spaces are developed. This expansion is completed at about LPI 4.0 when the 'mature' thickness of the lamina is reached, with a mean value of 248 μm. Between LPIs −4.25 and 0, the rate of expansion in thickness, dT/dt, is about 1.1 μm per day. The maximum rate of expansion of about 21 μm per day is reached at plastochron age 1.75. The middle part of the leaf becomes mature at LPI 4.0, or shortly thereafter. It is interesting to note that the most rapid rate of increase in lamina thickness occurs between LPIs 1 and 3. During this period of development seven to ten compact layers are established, cell division is rapidly declining and cell enlargement proceeds at a high rate.

It is difficult to say to what extent the results obtained for *Xanthium* leaves are representative of the course of leaf development in the herbaceous dicotyledons, or higher plants in general, since no other study of which the author is aware presents leaf growth data in a form which would allow

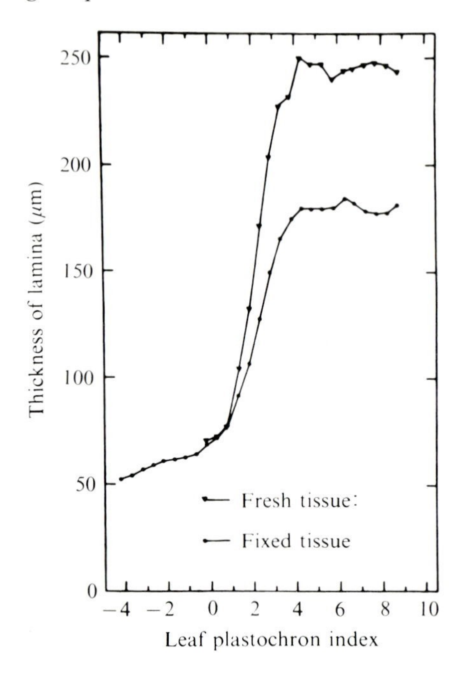

Fig. 15. Thickness of the lamina is plotted vs. LPI for fresh and for fixed and sectioned leaves.

estimates to be made of the time duration of various stages of development. Some authors (notably Avery, 1933) have cited leaf lengths corresponding to various stages of development, and some comparison can be made on this basis. Lamina initiation occurs in *Xanthium* at about LPI −4.8 when the leaf primordium is about 220 μm long. Avery reports that in *Nicotiana* the first indication of lamina formation occurs when the leaf primordium is approximately 600 μm long. Comparisons of other stages as well indicate that corresponding stages are found at greater leaf lengths in *Nicotiana* than in *Xanthium*. For instance, the period of rapid expansion in thickness of the lamina, with expansion of the palisade cells and development of intercellular spaces in the spongy parenchyma, occurs in *Xanthium* at about LPI 2.0, when the leaf is about 50 mm long. The corresponding stage in *Nicotiana* is at a leaf length of 80–100 mm. In both cases this corresponds to about one-third of the mature length of the leaf.

7

Differentiation of the laminar tissue

At least four different anatomical stages can be distinguished in the course of lamina development in *Xanthium pennsylvanicum*.

1. Through the activity of the marginal meristem, usually six basic layers of cells are established. They consist of adaxial and abaxial epidermal layers, two subepidermal layers, and the basic layers of the middle mesophyll (Fig. 16).

2. Around LPI 1.8, in the middle portion of the lamina, the plate meristem forms seven, eight or more compact layers by periclinal cell division (Fig. 17).

3. Starting with LPI 2.0 a rapid enlargement and differentiation of cells takes place. During this stage intercellular spaces are formed and the mesophyll tissue acquires its typical morphology (Fig. 18).

4. Finally, after LPI 4.0 the lamina becomes mature (Fig. 19).

Fig. 20 represents a series of cross sections through the lamina of the tip, middle and basal portions of the leaf. The relative units in parentheses indicate positions of these sections in the blade. At LPI −0.72 in all three portions, tip, base and middle part of the lamina, we find only six undifferentiated cell layers. The tip at LPI 0.68 has essentially eight compact layers and at LPI 1.78 it indicates a more advanced developmental stage, in which it displays a much thicker lamina, elongated palisade cells and intercellular spaces. In the corresponding middle and basal portions, however, very little change can be observed. At LPI 2.3 the lamina of the tip is mature, the middle portion has somewhat differentiated palisade cells with intercellular spaces in the middle mesophyll, but the basal portion still indicates almost no differentiation in its component six basic layers. The next stages of LPI 2.67 show a mature tip, the middle portion nearing maturity, and the basal portion with somewhat enlarged palisade cells and advanced intercellular spaces of the mesophyll tissue. Around LPI 4.2 the whole lamina displays completely differentiated tissues.

There is another way of approaching this developmental analysis by considering separately each portion with respect to progressively increasing plastochron age. The tip matures about two plastochrons sooner than the basal portion. It is evident that differentiation of the tissue in the lamina of the *Xanthium* leaf proceeds basipetally. This process of differentiation appears to be so gradual that no sharp boundary can be delineated in transition from one stage to another.

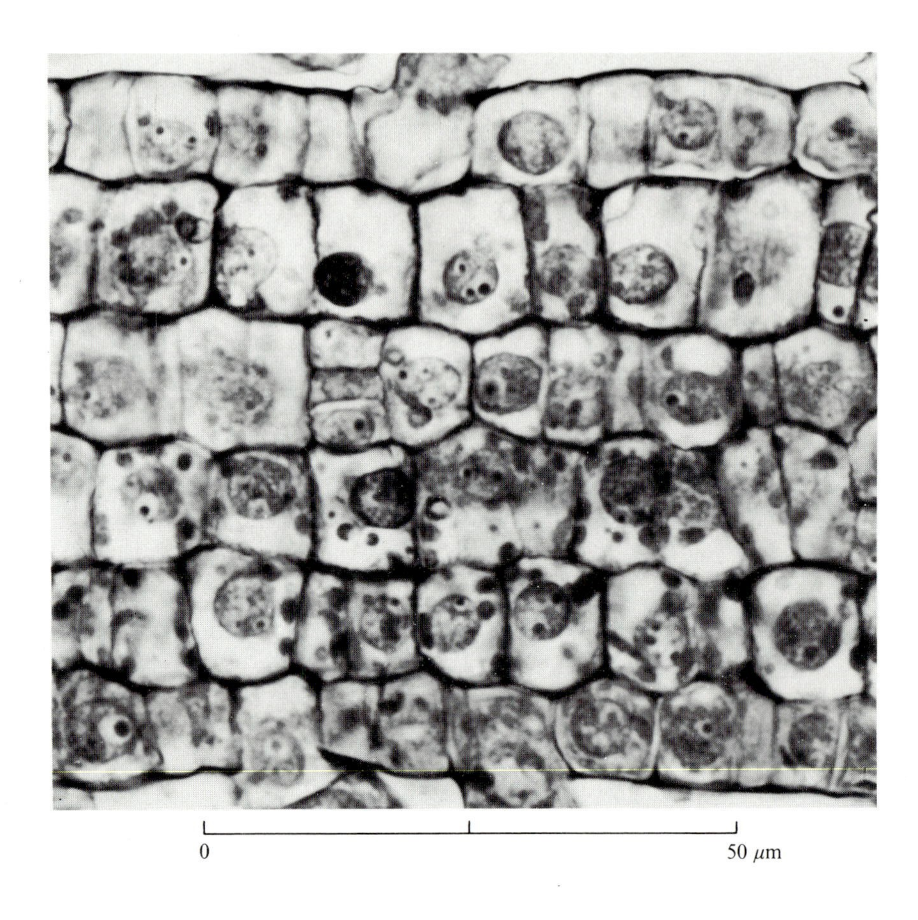

Fig. 16. Cross section of a *Xanthium* leaf at LPI −0.33 taken at mid-point between the tip and the base of the lamina. At this developmental stage the plate meristem consists of six compact layers of cells. The adaxial layer is on top.

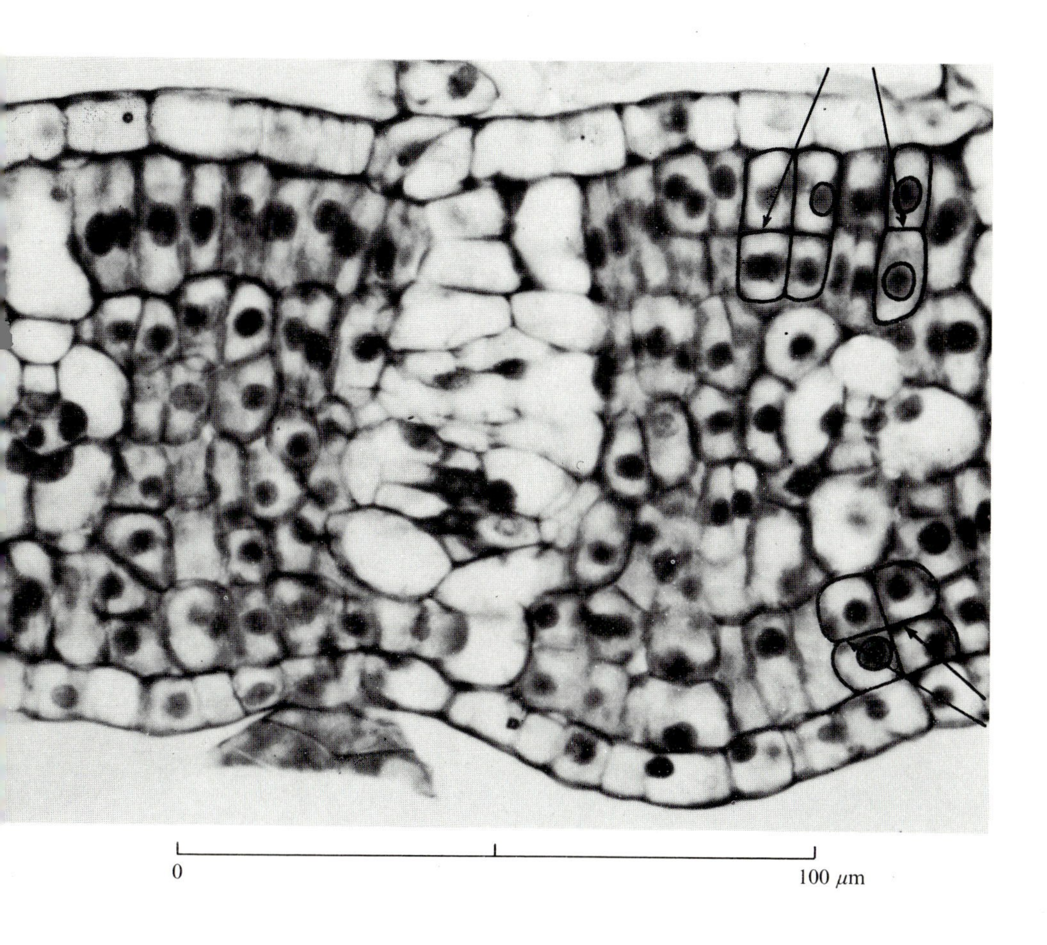

Fig. 17. Cross section of a more advanced developmental stage of a *Xanthium* leaf lamina at LPI 1.66 with 8–10 compact layers of cells. The increased number of cell layers originated from periclinal divisions indicated by arrows. Some enlargement of cells is noticeable at this stage.

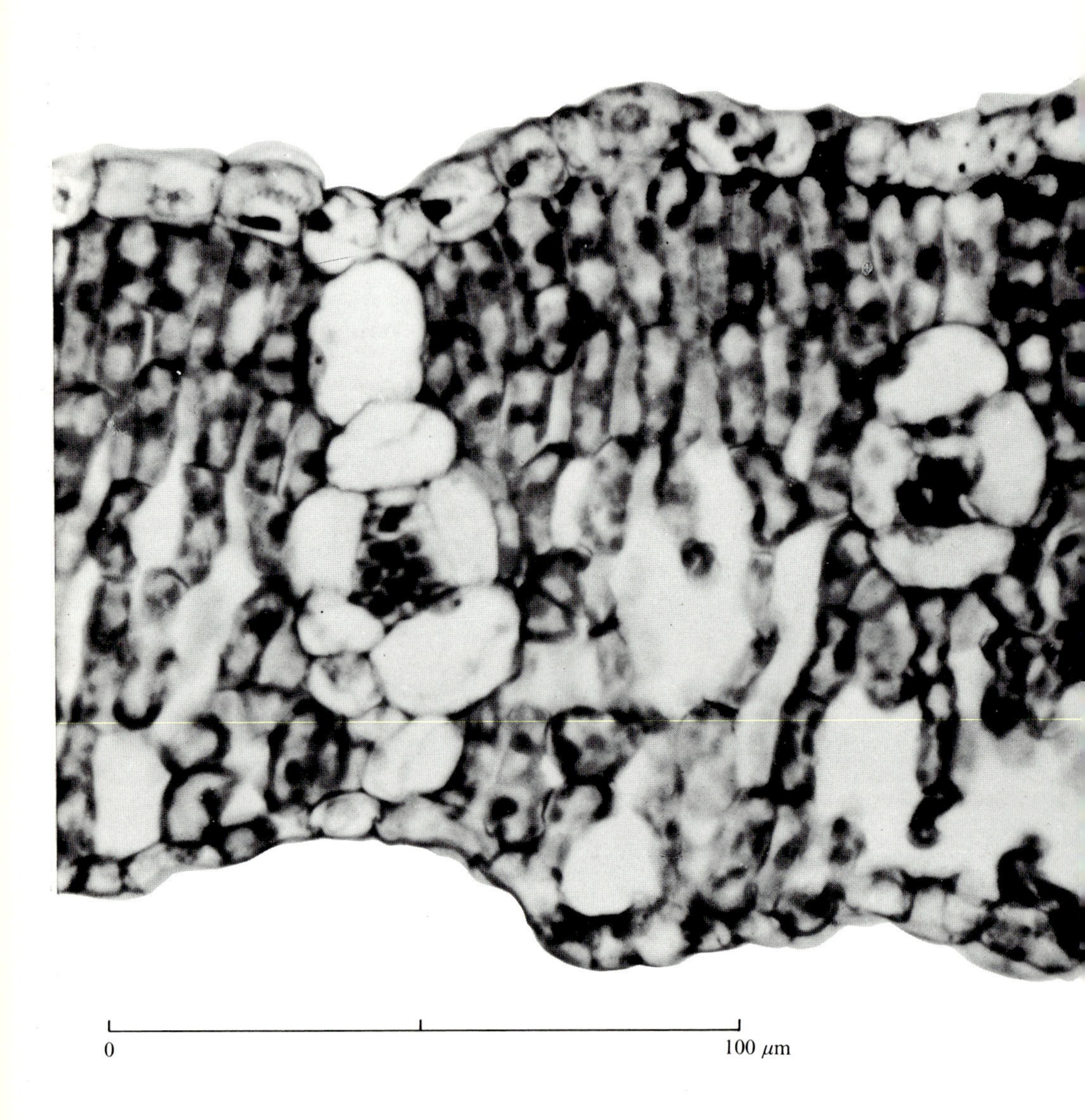

Fig. 18. Lamina development at LPI 2.6 is characterized by rapid cell enlargement, tissue differentiation and formation of intercellular spaces. Note still unseparated palisade cells.

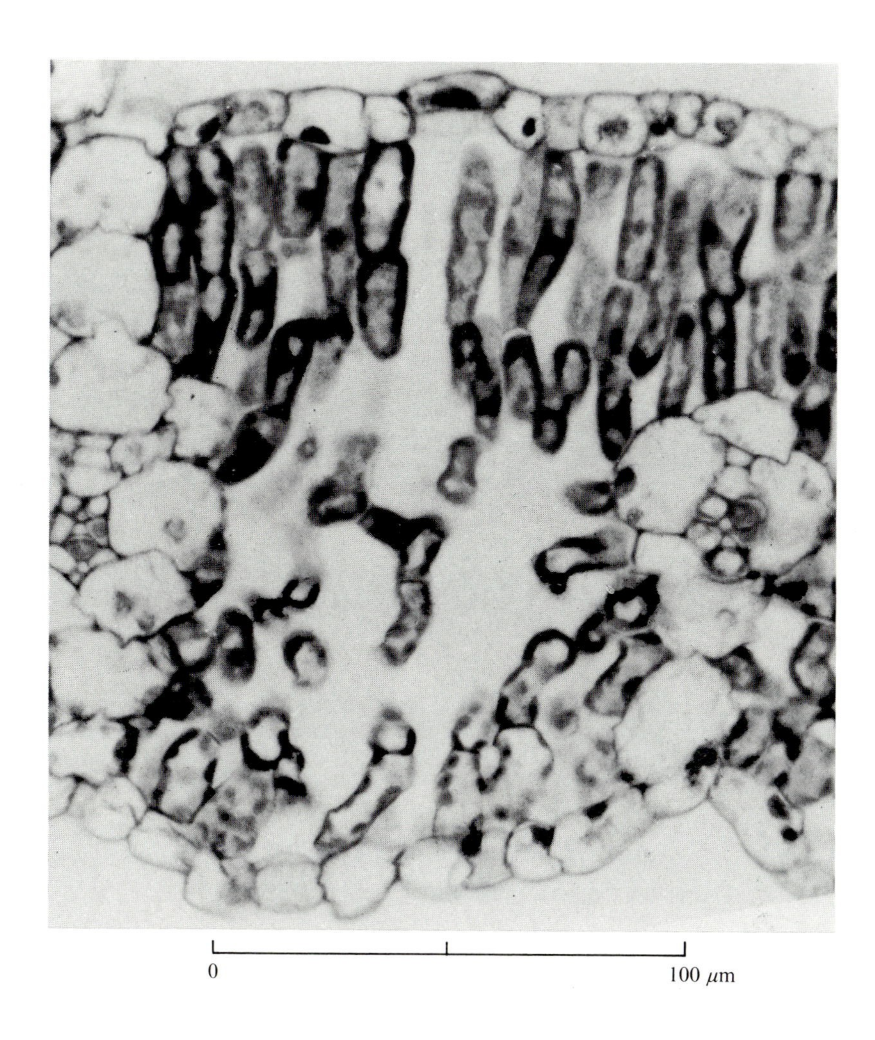

Fig. 19. Mature lamina of a *Xanthium* leaf at about LPI 7.0 has separated palisade parenchyma and a typical spongy mesophyll.

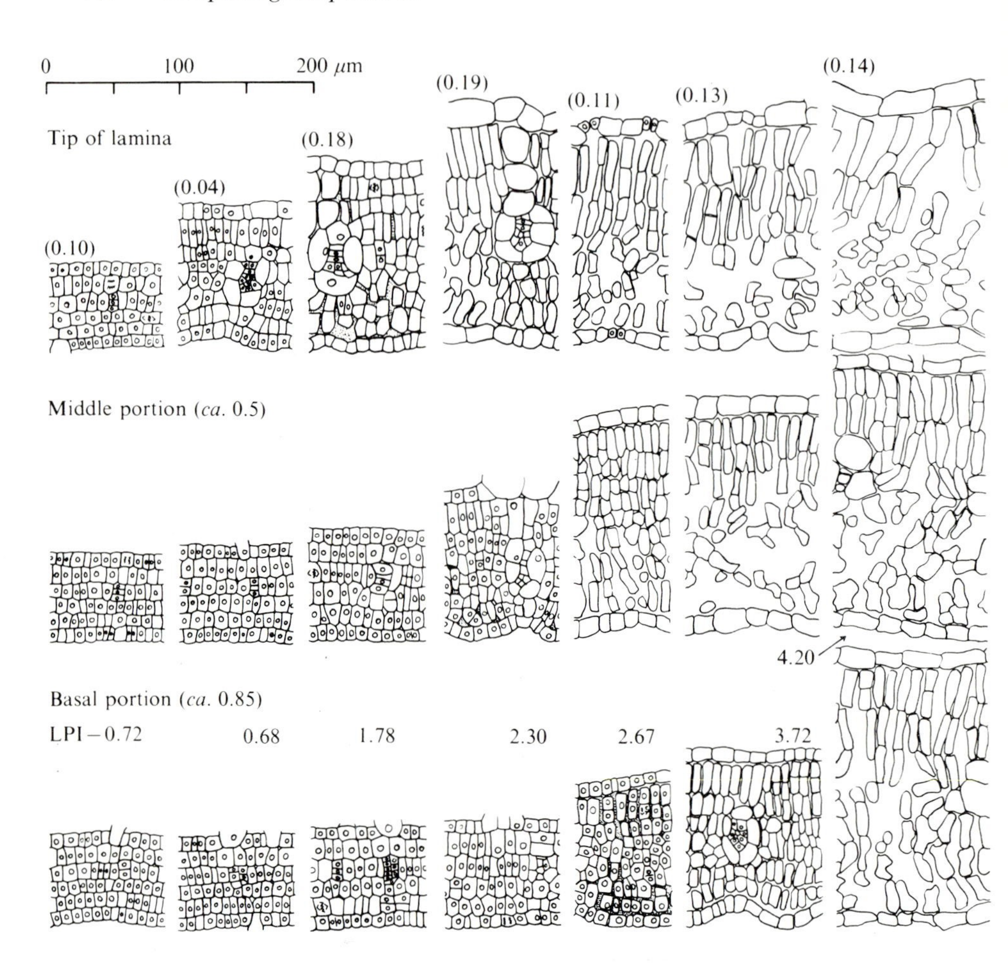

Fig. 20. Various stages of lamina development of the tip, middle and basal portions of the leaf representing basipetal differentiation of the tissue. Numbers in parentheses indicate relative position of leaf sections from the tip.

8

Leaf length

Leaf length has frequently been used in studies of plant development. It represents an important dimension of morphological change associated with growth. A careful analysis of leaf elongation frequently may reveal the intricate complexity of the pattern of development.

Lengths of the youngest leaf primordia can easily be estimated by counting serial sections from a transversely sectioned stem apex, starting with LPI -6.0. In the course of development leaf lengths can be measured from longitudinal sections or in living condition directly on the stem apex. The leaf primordium grows primarily by the mitotic activity of its apical meristem and by intercalary growth consisting of both cell division and enlargement. Leaf elongation can be represented by a sigmoid curve which is typical of many biological systems. Early leaf growth in *Xanthium* appears to be exponential as is evident from Figs. 21, 22, 23 and 24, and this condition is maintained to about LPI 2.5. The linear phase of elongation is found roughly between LPIs 2 and 5 with noticeable deceleration of growth in older plastochrons. The whole leaf stops elongating at LPI 7.0 or shortly thereafter, with an average length of 156 mm. The absolute rate of elongation, dL/dt, where L is the leaf length in mm and t is the time in days, represents an instantaneous rate of change in leaf length per day. The entire period of growth is represented by a bell-shaped curve. At young stages of development up to LPI -1.0 the increment is measured in fractions of a millimeter per day. The maximum rate of increase in leaf length between LPIs 3.0 and 4.0 is about 8 mm per day. Subsequent growth rates decrease rapidly and elongation stops completely after LPI 7.0.

From data on leaf elongation, it is possible to estimate the absolute rate of elongation (dL/dt) for the entire leaf. In addition, one would like to know whether each element of the leaf grows at a constant rate throughout the whole course of development, and, if not, what the growth distribution along the midrib of the leaf is. A series of marking experiments was undertaken (Maksymowych, 1962) in the expectation that the derived rates would provide some indication as to the pattern of elongation of each portion of the leaf. *Xanthium* leaves were marked with India ink at about 1 mm intervals along the midrib (Fig. 25) and were photographed on three successive days. Leaf 9 of a particular plastochron age was used only in one experiment and

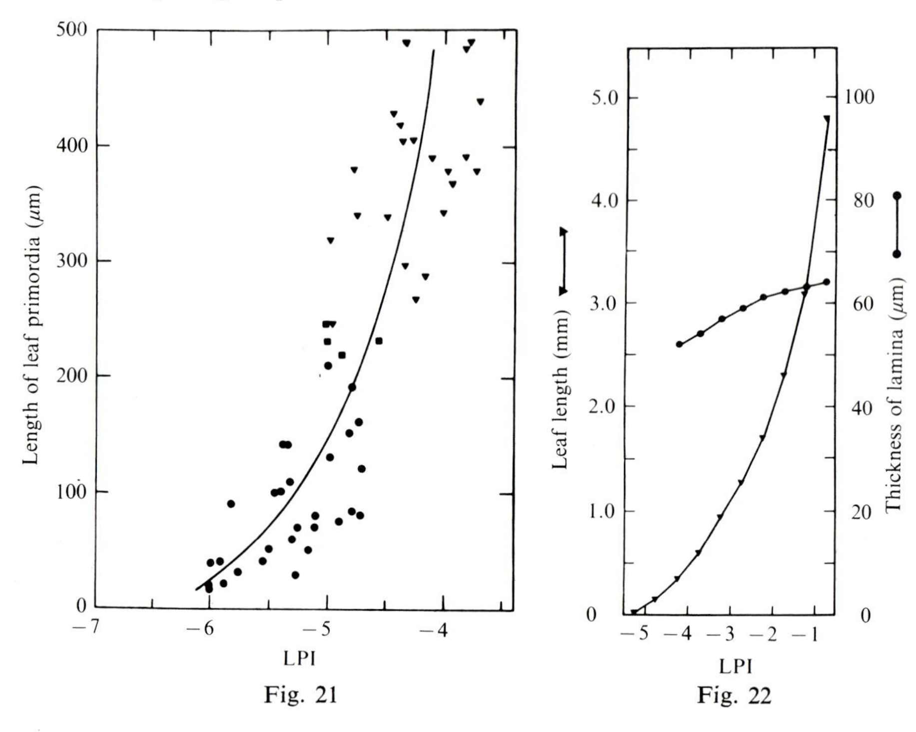

Fig. 21. Lengths of the two youngest leaf primordia of several shoots plotted vs. leaf plastochron index (LPI). Circles represent the youngest leaf primordia with no evidence of lamina initiation. Triangles represent the second youngest primordia with initiated laminae. The two squares around LPI −5.0 show the first indication of marginal activity.

Fig. 22. Lengths of young leaves and lamina thickness at early stages of development are plotted vs. LPI.

consequently yielded only one curve. This meant that in order to obtain another growth curve, for instance at LPI 4.5, leaf 9 from a different plant was used. The successive days on which the leaf was photographed, were designated as times −1, 0, and +1. It was assumed that during leaf elongation, the India ink marks would separate from each other to a different degree, depending upon their distance from the tip, thus establishing a growth pattern typical to each developmental age of the leaf. With cessation of growth of any portion of the leaf, the distance between two corresponding marks should remain unchanged.

Table 1 represents a method of analysis for leaf 9, at LPI 0.75. Distance of each mark from the tip of the leaf was measured separately for each day and designated as X_{t-1}, X_{t0} and X_{t+1} (Table 1, columns 1–3). These values were multiplied by the magnification factor in order to obtain distance in mm and they appear as X' values in columns 4–6.

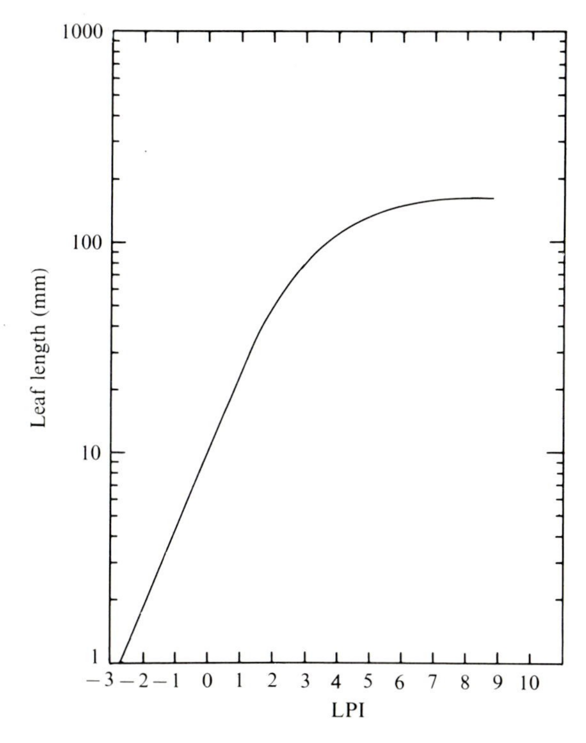

Fig. 23. Average leaf length plotted logarithmically vs. LPI. The early growth is exponential. The leaf is mature around LPI 7.0.

The rate of displacement of points, dX/dt, where X is the distance from the tip and t is time in days, was calculated with a 3-point formula for numerical differentiation. The dX/dt rates were plotted against their respective distances from the tip, which in the present example were the X'_{t_0} values. A curve was fitted by eye to the set of plotted points, and then smooth dX/dt values were read off the graph at equally spaced distances from the tip (Table 1, columns 8 and 9). From smooth dX/dt values, the relative elemental rates of elongation, $(d/dX)(dX/dt)$, were calculated with a 3-point derivative formula and then smoothed with a 7-point smoothing formula (Table 1, columns 10 and 11). The relative elemental rate of elongation is defined here as the rate of elongation of an infinitesimally small segment of the leaf at any distance from the tip, along the midrib, relative to its size. Since the plastochron index is linearly related to time, the time rates can easily be converted to plastochron rates, and vice versa, by simply multiplying the day rates by the duration of one LPI in days. The average duration of one plastochron in present studies is about 3.5 days.

Fig. 26 represents the rate of displacement of points, dX/dt (solid line),

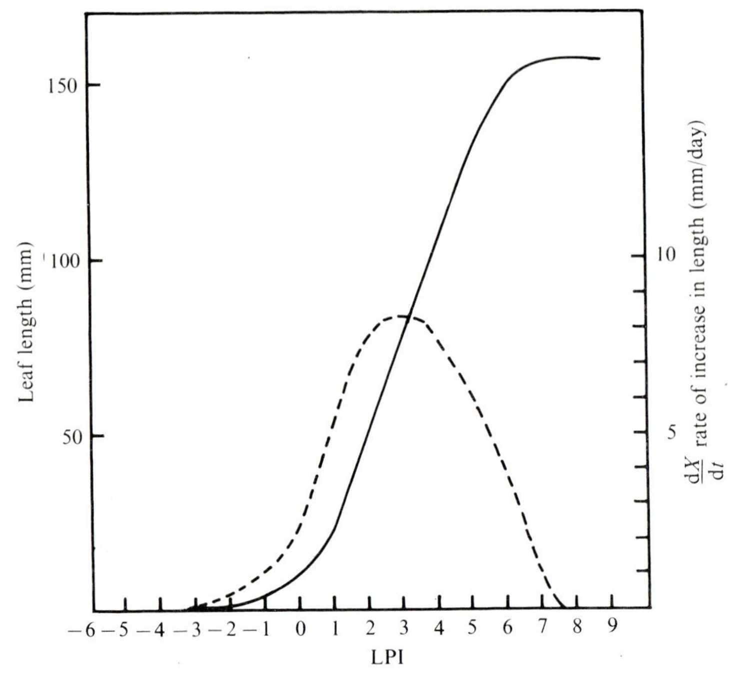

Fig. 24. Average leaf length and rate of increase in length plotted vs. LPI. The maximum rate of elongation is around LPI 3.0.

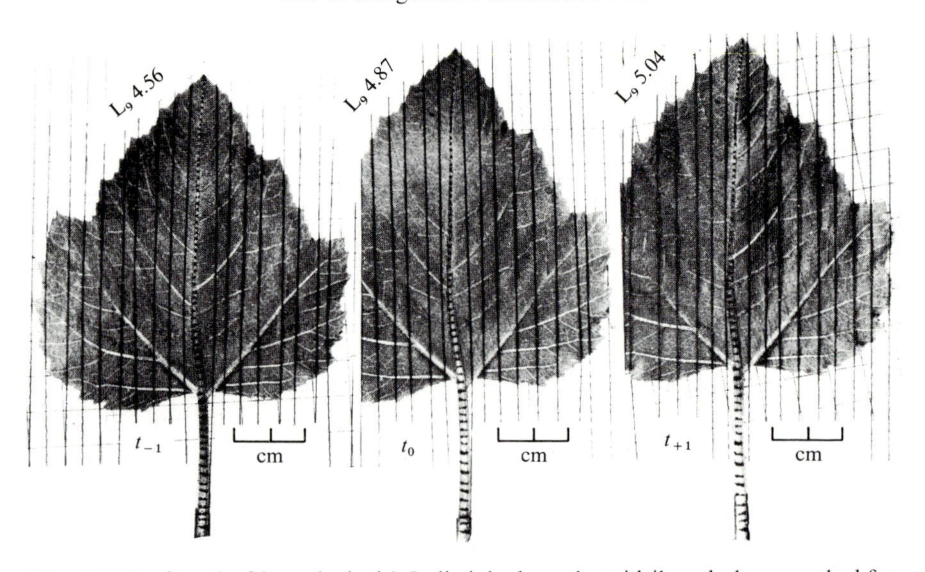

Fig. 25. *Xanthium* leaf 9 marked with India ink along the midrib and photographed for three consecutive days. L_9 4.56, L_9 4.87, and L_9 5.04 indicate plastochron ages of leaf 9; t_{-1}, t_0 and t_{+1} are designations for three successive days.

$X_{t_{-1}}$	X_{t_0}	$X_{t_{+1}}$	$X'_{t_{-1}}$	X'_{t_0}	$X'_{t_{+1}}$	$\frac{dX}{dt}$	X Distance from tip	Smooth $\frac{dX}{dt}$	3 pt. der. $\frac{d}{dX}\left(\frac{dX}{dt}\right)$	7 pt. smooth $\frac{d}{dX}\left(\frac{dX}{dt}\right)$
1	2	3	4	5	6	7	8	9	10	11
0	0	0	0.00	00.0	00.0	0.000	0.0	0.00	0.00	0.000
8	9	10	0.59	0.67	0.74	0.075	0.5	0.06	0.15	0.156
11	13	15	0.82	0.96	1.11	0.145	1.0	0.15	0.19	0.172
16	18	22	1.19	1.33	1.63	0.220		0.25	0.18	0.194
21	25	31	1.56	1.85	2.30	0.370	2.0	0.33	0.21	0.216
25	30	36	1.85	2.22	2.67	0.410		0.46	0.24	0.225
30	35	44	2.22	2.59	3.26	0.520	3.0	0.57	0.23	0.226
32	38	47	2.37	2.81	3.48	0.555		0.69	0.21	0.212
41	47	58	3.04	3.48	4.30	0.630	4.0	0.78	0.19	0.201
45	53	64	3.33	3.93	4.74	0.705		0.88	0.20	0.204
53	63	78	3.93	4.67	5.78	0.925	5.0	0.98	0.22	0.211
55	66	82	4.07	4.89	6.07	1.000		1.10	0.22	0.212
62	73	91	4.59	5.41	6.74	1.075	6.0	1.20	0.20	0.205
65	78	97	4.81	5.78	7.18	1.185		1.30	0.19	0.199
73	87	108	5.41	6.44	8.00	1.295	7.0	1.39	0.20	0.198
77	92	115	5.70	6.81	8.52	1.410		1.50	0.21	0.202
85	102	127	6.30	7.56	9.41	1.555	8.0	1.60	0.20	0.203
88	105	130	6.52	7.78	9.63	1.555		1.70	0.20	0.199
96	115	143	7.11	8.52	10.59	1.740	9.0	1.80	0.20	0.200
99	120	148	7.33	8.89	10.96	1.815		1.90	0.20	0.208
106	125	155	7.85	9.26	11.48	1.815	10.0	2.00	0.22	0.214
110	132	164	8.15	9.78	12.15	2.000		2.12	0.22	0.213
119	141	176	8.82	10.44	13.04	2.110	11.0	2.22	0.20	0.199
122	145	181	9.04	10.74	13.41	2.185		2.32	0.18	0.183
134	157	196	9.93	11.63	14.52	2.295	12.0	2.40	0.16	0.175
137	161	201	10.15	11.93	14.89	2.370		2.48	0.18	0.174
145	172	213	10.74	12.74	15.78	2.520	13.0	2.58	0.19	0.177
148	174	216	10.96	12.89	16.00	2.520		2.67	0.17	0.183
159	188	233	11.78	13.93	17.26	2.740	14.0	2.75	0.18	0.183
162	192	237	12.00	14.22	17.55	2.775		2.85	0.19	0.182
169	198	245	12.52	14.67	18.15	2.815	15.0	2.84	0.17	0.173
172	203	249	12.74	15.04	18.44	2.850		3.02	0.15	0.150
177	210	259	13.11	15.55	19:18	3.035				

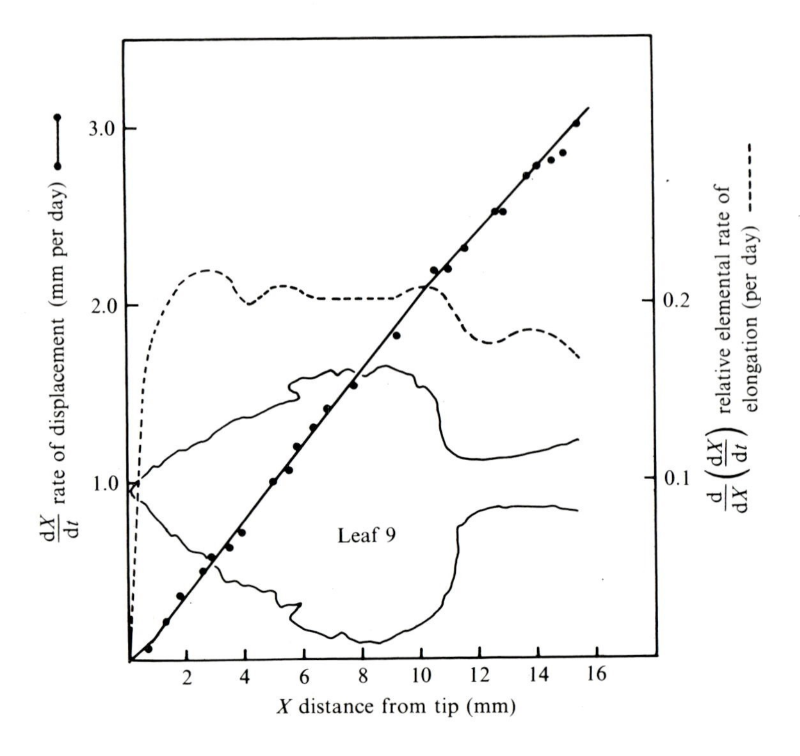

Fig. 26. Rate of displacement of marks and the relative elemental rate of elongation of a young leaf 9 at LPI 0.75 are plotted vs. distance from the tip.

and the relative elemental rate of elongation, $(d/dX)(dX/dt)$ (dashed line), for a young leaf of LPI 0.75 which is about 16 mm long, plotted against distance from the tip. With the exception of the very tip, the first curve appears to be a straight line up to 12 mm distance from the tip, indicating that the rate of displacement of points is constant. However, the main interest should be focused upon the second curve because it gives the relative rates of each element of the leaf. A general conclusion may be reached that, with the exception of the very tip, this young leaf was elongating at a constant relative elemental rate. It is believed that the wavy trend of the curve is due to variability of data, therefore probably not significant. Each segment of the leaf midrib indicates an increase of less than 20% per day.

More advanced stages of development are presented in Fig. 27. To illustrate a change of developmental pattern at increasing plastochrons a few more curves should be analyzed. The curves at LPIs 3.0 and 3.7 display a changed pattern of elongation. The relative elemental rates of elongation at the tip are low but increase in a basipetal direction, reaching a maximum

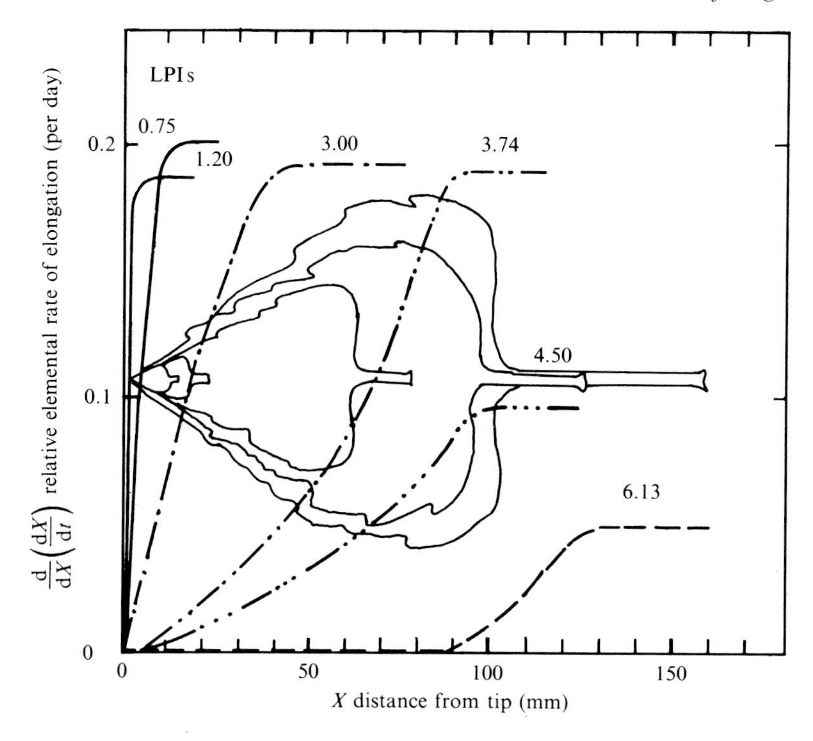

Fig. 27. A résumé graph for relative elemental rates of elongation of six leaves at various stages of development plotted vs. distance from the tip. The various curve profiles indicate a basipetal pattern of growth.

value of about 20% increase per day at their basal parts. This would indicate that the tissues at the tip of the leaf differentiate and mature ahead of the basal parts. If a leaf is sectioned at LPI 2.5 and its anatomy studied, the tip of this leaf is mature, the middle portion shows a high degree of tissue differentiation and the basal lobes of the lamina are primarily composed of meristematic tissue. The correlation of the elemental rate analysis seems to fit well into the general developmental picture.

The leaf of LPI 6.1 (Fig. 27) is about 165 mm long and, anatomically considered, it has a mature blade. The relative elemental rate of elongation for this leaf, up to 100 mm distance from the tip, is 0, but then it constantly increases. Evidently, the lamina of this leaf does not elongate at this stage but only the basal part of the petiole grows at a rate of 5% per day. Apparently the growth of this leaf is due to slight increase in petiole length.

It is evident that the relative elemental rate of elongation is not constant throughout the whole leaf length, but changes with increasing plastochron age. This means that each element of the leaf will elongate at a different value of $(\mathrm{d}/\mathrm{d}X)(\mathrm{d}X/\mathrm{d}t)$ depending upon its position with respect to the distance from the tip and upon the age of the leaf. There is a definite trend in the magnitude of these rates which decline first at the tip, thereafter progressively diminishing toward the base insertion of the petiole. However, the elements of young leaves, up to LPI 0.75, elongate at a constant rate. Another implication may be that cell division and cell elongation would stop first at the tip, but still proceed at the base of the leaf. This is in agreement with reports by Von Papen (1935) and Maksymowych (1959), where evidence was presented that leaf tissues displayed a definite trend of basipetal differentiation. For instance, at LPI 2.3 the tip of the lamina may be mature, the middle portion has somewhat differentiated palisade cells and intercellular spaces of the middle mesophyll, but the basal portion still indicated almost no differentiation of its component six basic layers of cells.

9

Leaf area

The leaf is a photosynthetic organ. It converts the light energy into the chemical energy which is stored in the form of sugar or starch. The amount of photosynthesis under usual conditions is a function of the leaf area. It would be of interest, therefore, to follow the area growth as a function of time and to correlate it with other developmental parameters.

Even though the lamina is initiated at LPI −4.8, it becomes practical to start measurements on leaves older than LPI −1.0 when enough surface area is established. It can be seen from Fig. 28 that the increase in leaf surface area follows a typical sigmoid curve. Since the logarithmic plot of the area vs. LPI yields a straight line (Fig. 29), it can also be concluded that this increase in the early plastochrons is exponential. The mature blade is established around LPI 6.0.

The absolute rate of growth (dA/dt), where A represents the leaf area in cm^2 and t is time in days, is only 0.25 cm^2 per day at zero LPI. The maximum rate of about 10 cm^2 per day is reached between LPIs 3.5 and 4.0 with a rapid decrease of growth rate thereafter. As indicated above, the absolute rate of growth represents an increase in surface growth of the whole leaf per unit of time. One would like to know whether various parts of the lamina expand at the same or different rates during the course of development. To get this information, relative rates $(1/A)(dA/dt)$ of small areas delineated by veins were calculated at various distances from the tip of LPIs 0.74, 2.64 and 4.18. Numbers listed in various parts of the lamina (Fig. 30) represent the relative rates of increase in surface area of the designated portions. If multiplied by 100, these numbers indicate a percent increase in cm^2 per day. Even though these data are somewhat variable, nevertheless some pertinent conclusions can be reached. Various parts of the lamina expand at different rates, depending upon their distance from the tip and the age of the leaf. The relative rate of growth is lowest at the tip of the leaf and increases in the basal direction, thus conforming to the basipetal pattern of expansion. This conclusion is in agreement with the marking experiments on the midrib of *Xanthium* leaf in which the relative elemental rates of increase in length $(d/dX)(dX/dt)$ were estimated and also with the basipetal pattern of tissue differentiation. Erickson (1966) studied the surface growth of *Xanthium* leaf by a method which consists of evaluating the divergence of

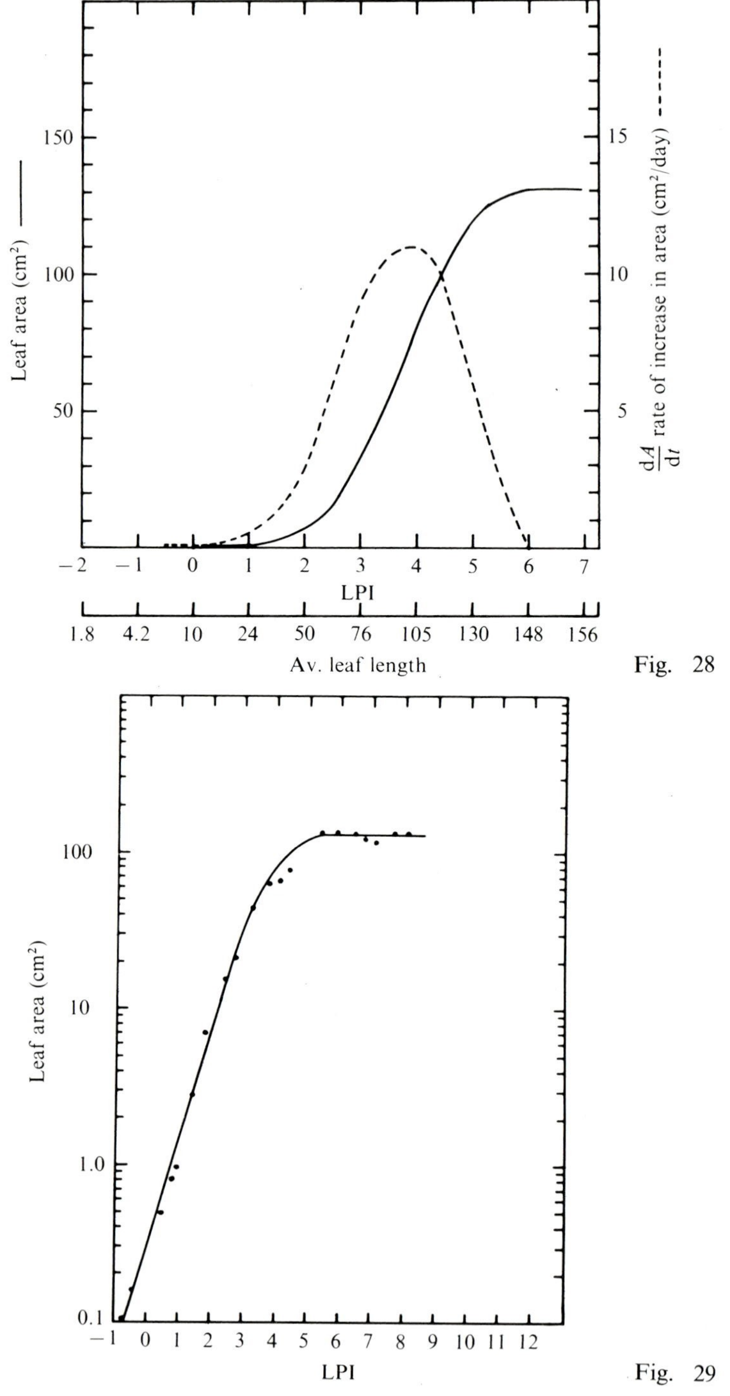

Figs. 28 and 29. Leaf area in cm^2 on the left ordinate and rate of increase in area on the right ordinate plotted vs. LPI. Average leaf lengths are indicated at corresponding LPI values. Maximum rate of expansion takes place between LPIs 3 and 4. Lamina growth stops after LPI 6.0. The same leaf area is plotted logarithmically in Fig. 29. The early expansion of the lamina appears to be exponential.

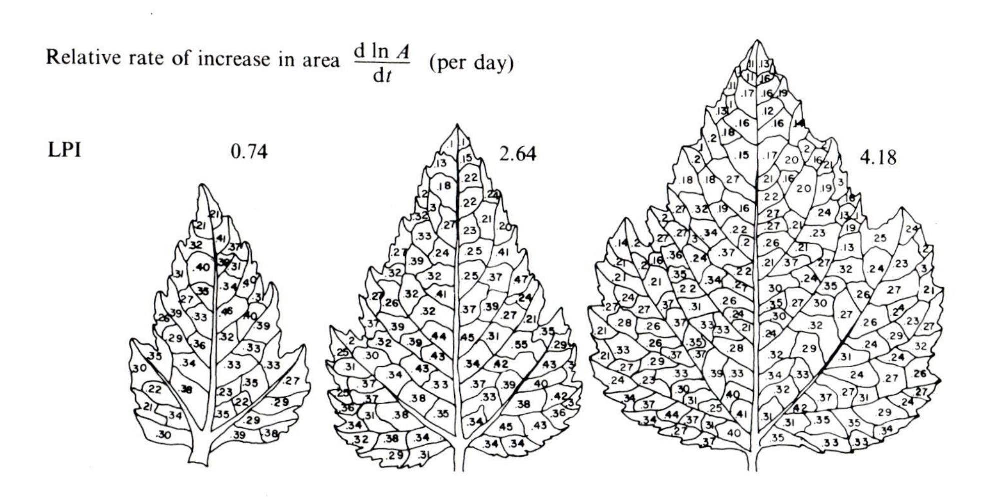

Fig. 30. Relative growth rates of small leaf areas estimated at LPIs 0.74, 2.64 and 4.18. Various parts of the lamina expand at different rates depending upon their distance from the tip and the age of the leaf conforming to a basipetal pattern of growth.

velocity by vector analysis. This method has been adapted for solution by numerical methods, and has been programmed for an electronic computer. A contour map of a leaf at LPI 2.5 is illustrated in Fig. 31. Erickson concluded that leaf expansion is largely isotropic, though the margins are somewhat anisotropic. The direction of maximum expansion in the anisotropic growth was approximately parallel with the margin. In addition to a general basipetal pattern of expansion, Erickson demonstrated that there was a general tendency for the center of the leaf to have higher rates than the margin which can also be seen in Fig. 30. The previously discussed analysis presented in relative rates of expansion of various parts of the lamina is loosely comparable with Erickson's relative elemental rate analysis. Higher rates in the basal parts of the lamina in Erickson's analysis could be attributed to differences in environmental conditions.

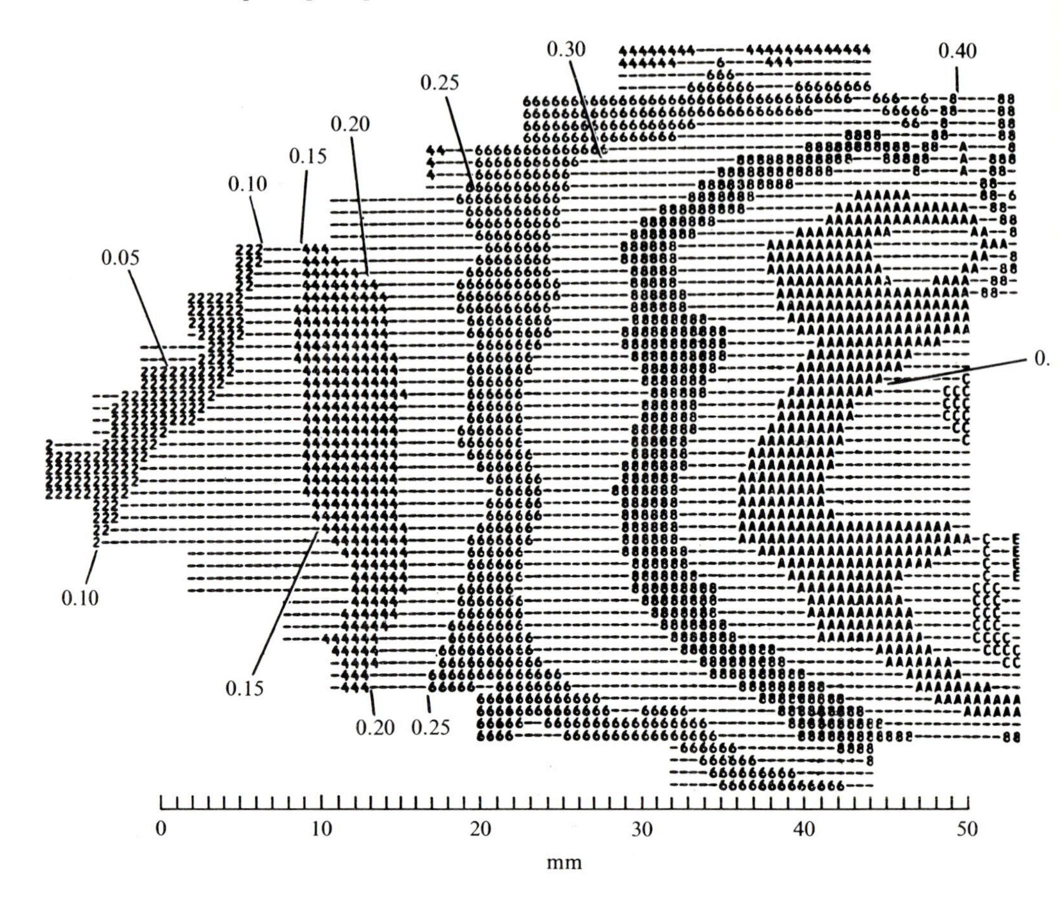

Fig. 31. Contour map of *Xanthium* leaf at LPI 2.5 in which digits and letters represent classes of relative elemental rate: 2, 0.05 to 0.10 per day; 4, 0.15 to 0.20 per day; A, 0.45 to 0.50 per day, etc. Hyphens represent odd numbers. (From Erickson, 1966, *Jour. Exp. Bot.*, **17**, by permission of the Clarendon Press, Oxford.)

10

Cell division

Cell division can be studied by estimating the number of cells at various stages of leaf development (Maksymowych, 1959 and 1963). This approach involves maceration of tissues into separate cells and making cell counts on a hemocytometer. From the sample count the total number of cells of the leaf at any desired developmental stage is then estimated. The advantage of this approach is that rates of cell division can be calculated because the time factor of increase in the number of cells is taken into consideration. The most conventional method in studies of cell division is to determine the proportion of mitoses in a developing system, 'the mitotic index'. This index, however, yields data unsuitable for the estimation of rates of cell division because the time element is absent. Further inaccuracies arise due to difficulties in making a precise assessment of the early prophases and late telophases. Nevertheless, the mitotic index can be used as a relative measure of the mitotic activity of a developing system.

A widely used method for measuring the various parts of the cell division cycle takes advantage of the fact that labeled thymidine, a specific precursor of DNA, is readily incorporated into nuclear DNA of actively dividing cells. The autoradiographic method is especially suitable for estimation of the various parts of the cell cycle (G_1, S, G_2 and M).

Fig. 32 represents a semilogarithmic plot of the number of cells vs. the plastochron age of the whole leaf 9 and its petiole. A straight line was fitted by the least squares method calculated over the logarithms of mean values compiled over 0.5 plastochron intervals. The number of cells increases exponentially up to about LPI 2.5, and levels off rapidly thereafter. This indicates that after LPI 3.0, the cells of the lamina cease dividing. The mean number of cells in a mature leaf 9 was estimated to be about 116 million, and about 156 million in leaf 13.

If the assumption is made that the number of cells in the early development increases exponentially and could be expressed by equation (1), the average generation time, frequency of cell division and the number of generations of cells can be estimated from this equation.

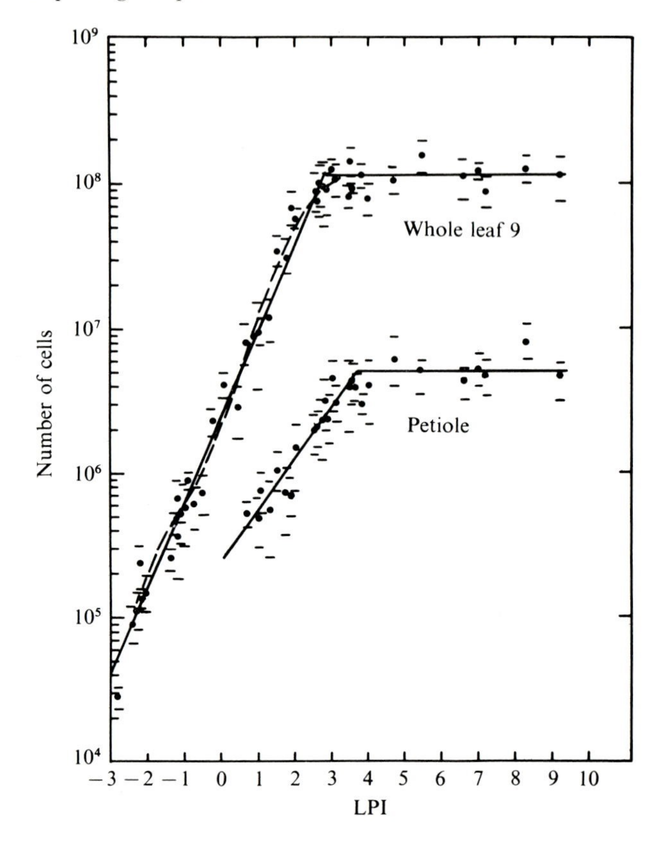

Fig. 32. Number of cells of the whole leaf 9 with its petiole plotted logarithmically vs. LPI. The increase in cell number in the early plastochrons appears to be exponential with cell division ceasing at LPI 3.0. Mature leaf has about 116 million cells.

$$N = N_0 \cdot 2^{rt}; \tag{1}$$

$n = r \cdot t$; $a = 1/r$; $a = t/n$; $n = t/a$;

N = total number of cells;

N_0 = initial number of cells;

r = frequency of cell division per day;

t = time in days;

n = number of generations;

a = generation time in days; or doubling time.

At LPI −2.0 the average number of cells is 150000; and at LPI 2.0 it is about 46 million. Doubling time can be estimated during this exponential period of development. Converting equation (1) to its logarithmic form:

$\log N = \log N_0 + (\log 2)\, rt,$
$(\log 2)\, r = \log N - \log N_0 / t = \text{slope},$
$(\log 2)\, r = 7.66276 - 5.20412/17.6 = 0.13969\ (\text{slope}),$
$r = \text{slope}/\log 2 = 0.13969/0.30103 = 0.464 =$ frequency of cell division per day,
$a = 1/r = 1/0.464 = 2.15 =$ doubling time in days.

Assuming that cell division started with a single cell and that $N = 116$ million cells, one can calculate the approximate number of generations, which is essentially the number of cell cycles occurring between the initiation of the primordium on the stem apex and the cessation of cell division.

$$N = N_0 \cdot 2^n, \qquad (2)$$
$\log N = (\log 2) \cdot n,$
$n = \log N / \log 2,$
$n = 8.06446/0.30103 = 26.8,$ or about 27 generations.

It can be concluded from the above consideration that after approximately every second day the number of cells in a *Xanthium* leaf will double. On the average, about 27 generations of cells are involved in the process of leaf formation from the time the primordium is initiated until cessation of cell division. Based upon the autoradiographic studies (Hearney, 1965), the total cell division cycle (generation time) in *Xanthium* leaves at LPI 0.5 is about 22 hours. The DNA synthesis period (S) is about 15.7 hours and the duration of mitosis is about 1.7 hours. The mitotic index can be estimated from the duration of mitosis divided by the duration of cell cycle. For a young *Xanthium* leaf, the average mitotic index is about 7.7%.

The total cell cycle can also be estimated from the percent of thymidine labeled nuclei plotted as a function of growth in the radioisotopic solution. Leveling of the curve after about 22 hours of growth (Fig. 47) would also indicate a significantly shorter generation time than the average doubling time of about 2 days derived from the regression line in Fig. 32. This difference can be attributed to the fact that not all cells are undergoing division, as is theoretically assumed in the application of the exponential equations (1) and (2). Obviously, some cells may be in a process of differentiation or have already differentiated into specialized tissues. The fact that they are included in computations, together with dividing cells, is responsible for the discrepancy. To obtain a generation time of about 22 hours, the duplication coefficient of the exponential equation would have to be changed from 2 to approximately 1.45. This would indicate that about 28% of cells in a cell population stop dividing to enter a process of differentiation. It should be understood, therefore, that the generation time, based on autoradiographic data for dividing cells only, is about 22 hours, whereas the time period for

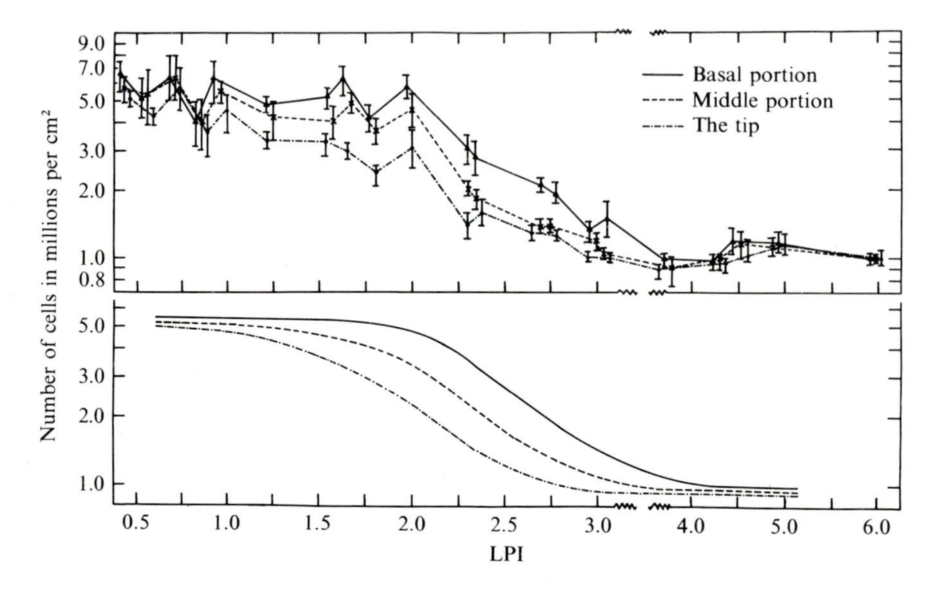

Fig. 33. Number of cells in millions per cm^2 of the tip, middle and basal portions of the lamina plotted vs. LPI. The lower graph is a smooth version of the upper plot to illustrate the basipetal trend of cell division. Cell division stops first at the tip and last in the basal parts of the leaf.

the doubling of cell population, designated as doubling time and calculated from the regression line, is about 2 days.

The next question one would like to pose is whether there is a constant pattern of cell division with respect to different parts of the leaf, whether all cells stop dividing at the same time in each portion of the lamina, or whether there is a different pattern of distribution. Fig. 33 represents a semilogarithmic plot of the number of cells per cm^2 of the tip, middle and basal portions of the lamina plotted vs. plastochron age of the leaf. These data were obtained from cell counts of the macerated leaf discs at various stages of development. Judging by the confidence intervals, there is no significant difference in the numbers of cells among the tip, middle and basal portions of the blade at early plastochrons, between LPIs 0.5 and 1.0 and also after LPI 3.0. In the early stages of development where the tissue is meristematic, there are on the average 5 million cells per cm^2. In final stages of development (LPI 4.0) this number is reduced to about 1 million cells per cm^2. This five-fold decrease indicates that cells have ceased dividing and have also completed their enlargement. There is a highly significant difference among the tip, middle and basal parts of the lamina between LPIs 1.5 and 2.5 with respect to the number of cells in each disc. It can be inferred that cells at the base of the leaf are in a meristematic condition and still may divide. In the middle and at the tip, cells are already in the process of elongation and cell division in all probability may have ceased there. It is difficult to indicate precisely the stage where cell division will stop at the tip, or any other portion of the blade, since the decline in cell number per unit area may be due to both the decreasing rates of cell division and cellular enlargement. Although it would be difficult to separate these two processes, there is some justification for drawing a general conclusion that the pattern of cell division in the *Xanthium* leaf is not uniform throughout the lamina at any one time. Cell divisions cease first at the tip and last at the basal lobes of the leaf, indicating a distinct basipetal trend.

11

Cell enlargement and differentiation

The leaf in its growth is a three-dimensional structure. It elongates, expands in area, and also grows in thickness. These manifestations of growth are due to cell division and cell enlargement. Cell division has already been discussed and consequently one would like to focus the attention on cell enlargement since this process is intrinsically associated with cellular differentiation. The upper epidermis and palisade mesophyll cells have been chosen in studies of cell growth because of the geometric regularity of these cells and also because of the importance of the relationship of the two tissues to each other during leaf development. The analysis of growth was carried on in terms of the absolute and relative rates during the entire period of development. Data and procedures for the calculations of rates are presented in Tables 2 and 3.

The meristematic palisade cell around LPI −1.0, on the average, is 10.4 μm high and 8.4 μm in diameter. As can be seen in Figs. 34 and 36, a rapid increase in height occurs between LPIs 1.0 and 4.0, with a maximum rate at LPI 3.0. The average mature cell is about 41 μm high. Considering the horizontal plane, which represents the cell diameter, it is interesting to note the drop in diameter between LPIs 1.0 and 2.0. It can be inferred that at this plastochron age the rate of anticlinal cell division has increased and may be in synchrony with the rapidly expanding epidermis (Figs. 35 and 36). The average diameter of a mature palisade cell is about 14 μm. Expansion in the vertical and horizontal planes seems to be synchronized and stops around LPI 5.0.

The embryonic epidermal cell (Fig. 35) is similar in size to its neighboring palisade parenchyma. It is 9.2 μm in height and elongates only a little in the vertical plane, reaching a height of 14.2 μm at LPI 5.0. The area of an embryonic epidermal cell, as visualized in a paradermal section, is about 50 μm^2, and the most rapid period of expansion is at LPI 3.0. At this plastochron age the absolute rate, dA/dt, has reached a maximum of about 125 μm^2 per day. Expansion in both planes seems to stop at the same plastochron age. The epidermal cell increases appreciably in area, but only insignificantly in its height, which is exactly the reverse polarity of growth shown by the palisade parenchyma.

One of the important aspects of development is the expansion of the

Table 2. Enlargement of palisade cells

LPI	Vertical plane (Height) H					Horizontal plane (Area) A						Volume V		
Mid-point of class interval	No. cells measured per LPI	Average cell length H (μm)	Smooth values cell length H (μm)	$\frac{dH}{dt}$ (μm per day)	$\frac{1}{H}\left(\frac{dH}{dt}\right)$ (per day)	No. cells measured per LPI	Average cell diameter X (μm)	Smooth values of cell diameter X (μm)	Area A (μm^2)	$\frac{dA}{dt}$ (μm^2 per day)	$\frac{1}{A}\left(\frac{dA}{dt}\right)$ (per day)	Cell volume (μm^3)	$\frac{dV}{dt}$ (μm per day)	$\frac{1}{V}\left(\frac{dV}{dt}\right)$ (per day)
−2.5	64	10.0[a]	10.4			64	8.2	8.4	55.4			576		
−2.0			10.4	0.00	0.000			8.4	55.4			576		
−1.5	40	10.6[b]	10.4	0.03	0.003	40	8.7	8.4	55.4			576		
−1.0			10.5	0.11	0.011			8.4	55.4			576		
−0.5	55	11.3	10.8	0.37	0.034	55	8.3	8.4	55.4	0.0	0.0	576	0.0	
0.0			11.8	0.63	0.053			8.4	55.4	−0.7	−0.013	576	30.8	0.054
+0.5	90	13.3	13.0	0.9	0.070	90	8.7	8.2	55.4	−3.5	−0.067	684	19.7	0.029
1.0			15.0	1.3	0.084			7.4	43.0	−3.4	−0.080	645	6.9	0.011
1.5	190	17.5	17.4	1.5	0.089	190	6.7	7.2	40.7	−0.6	−0.015	708	52.9	0.075
2.0			20.4	1.9	0.095			7.2	40.7	+0.6	+0.015	830	95.1	0.115
2.5	186	23.1	24.2	2.5	0.103	186	7.6	7.4	43.0	5.7	0.133	1041	270.0	0.259
3.0			29.2	3.0	0.102			8.8	60.8	12.9	0.211	1775	574.3	0.323
3.5	210	37.7	34.6	2.7	0.077	210	11.5	10.6	88.2	16.0	0.181	3051	792.3	0.206
4.0			38.6	1.6	0.040			12.2	116.8	13.9	0.119	4508	685.4	0.154
4.5	104	39.2	40.1	0.63	0.015	104	12.7	13.2	136.8	9.3	0.068	5485	454.8	0.083
5.0			40.8	0.20	0.005			13.7	149.5	4.9	0.033	6100	227.9	0.037
5.5	180	40.4[c]	40.8	0.00	0.000	180	13.5	14.0	153.9	1.2	0.008	6279	51.5	0.007
6.0			40.8					14.0	153.9		0.000	6279	0.0	0.000
6.5	196	41.1	40.8			196	13.2	14.0	153.9			6279		
7.0			40.8					14.0	153.9			6279		
7.5	30	40.6	40.8			30	13.8	14.0	153.9					
8.0			40.8					14.0	153.9					
8.5	110	41.9	40.8			110	13.8	14.0	153.9					

[a] $S_{\bar{x}} = 0.1563$; 95% C.I. = ±0.38   [b] $S_{\bar{x}} = 0.1471$; 95% C.I. = ±0.36   [c] $S_{\bar{x}} = 0.4910$; 95% C.I. = ±1.21

Table 3. Enlargement of upper epidermal cells

LPI	Horizontal plane (Area) A					Vertical plane (Height) H					Volume V		
Midpoint of class interval	No. cells measured per LPI	Average cell area A (μm^2)	Smooth values cell area A (μm^2)	$\frac{dA}{dt}$ (μm^2 per day)	$\frac{1}{A}\left(\frac{dA}{dt}\right)$ (per day)	No. cells measured per LPI	Average cell height H (μm)	Smooth values cell height H (μm)	$\frac{dH}{dt}$ (μm per day)	$\frac{1}{H}\left(\frac{dH}{dt}\right)$ (per day)	Cell volume (μm^3)	$\frac{dV}{dt}$ (μm^3 per day)	$\frac{1}{V}\left(\frac{dV}{dt}\right)$ (per day)
−2.5	60	52	50			60	9.2	9.2			460		
−2.0			50					9.2			460	0	
−1.5	40	52	50	0.0	0.000	40	9.3	9.2			460	0	0.000
−1.0			50	0.6	0.011			9.2			460	5	0.011
−0.5	56	49	52	1.4	0.029	56	9.3	9.2	0.00	0.000	478	13	0.027
0.0			55	2.3	0.043			9.2	0.00	0.003	506	23	0.045
+0.5	90	69	60	7.1	0.120	90	9.6	9.3	0.06	0.006	558	70	0.126
1.0			80	15.7	0.197			9.4	0.14	0.015	752	163	0.216
1.5	276	130	115	31.4	0.274	186	10.2	9.8	0.26	0.026	1127	344	0.305
2.0			190	64.3	0.337			10.3	0.29	0.028	1957	727	0.371
2.5	185	331	340	111.4	0.328	187	11.0	10.8	0.37	0.034	3672	1363	0.371
3.0			580	125.7	0.217			11.6	0.43	0.037	6728	1892	0.251
3.5	110	865	780	85.7	0.111	210	12.6	12.3	0.40	0.032	9594	1346	0.140
4.0			880	38.6	0.043			13.0	0.37	0.029	11440	813	0.071
4.5	220	917	915	14.3	0.017	104	12.5	13.6	0.29	0.021	12440	478	0.038
5.0			930	7.1	0.009			14.1	0.17	0.012	13113	259	0.020
5.5	60	944	940	2.9	0.003	166	13.7	14.2	0.03	0.002	13348	67	0.005
6.0			940	0.0	0.000			14.2	0.00	0.000	13348	0	0.000
6.5	266	921	940			152	14.3	14.2			13348		
7.0			940					14.2			13348		
7.5	37	924	940			16	14.2	14.2			13348		
8.0			940					14.2			13348		
8.5	110	986	940			83	14.3	14.2			13348		

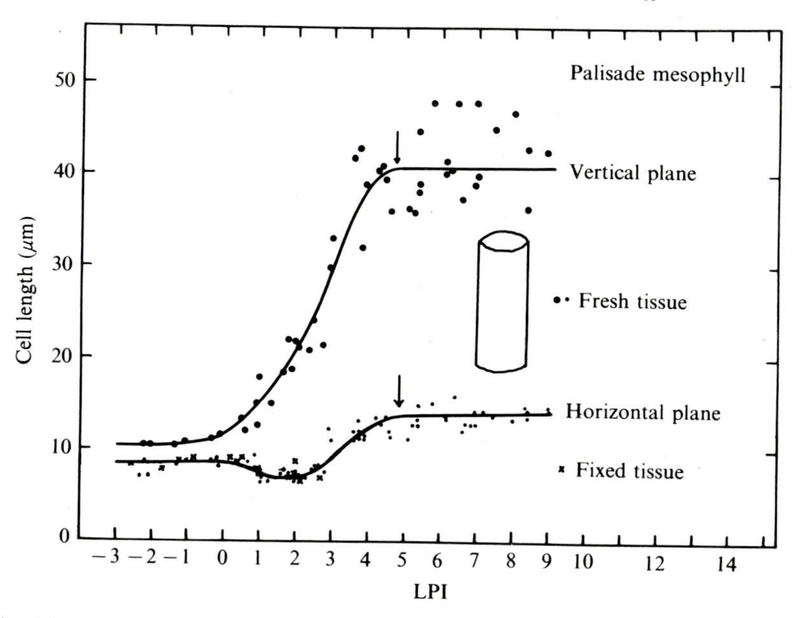

Fig. 34. Cell length of the palisade mesophyll plotted as a function of leaf plastochron age. Vertical plane is synonymous with cell height. Horizontal plane which is parallel to the lamina is synonymous with cell diameter. Growth in height is dominant and only slight increase in cell diameter is noticeable.

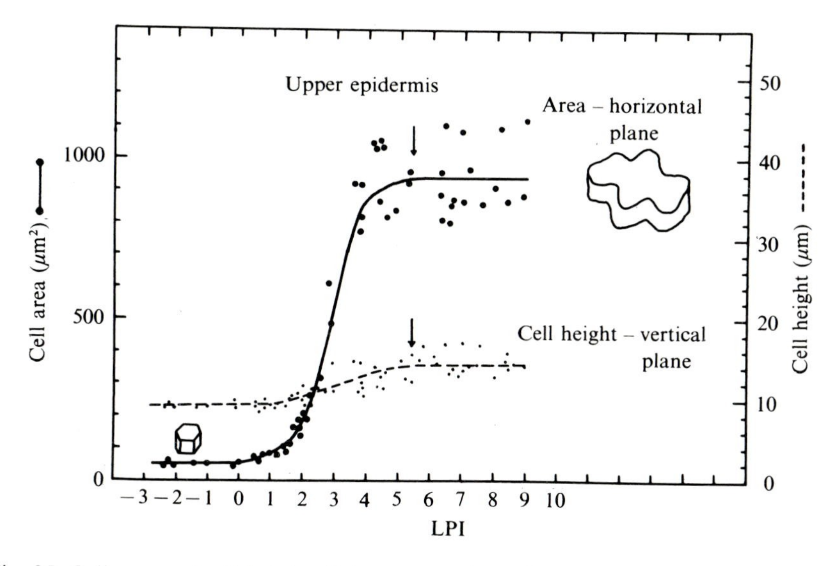

Fig. 35. Cell area and cell height of the upper epidermis plotted vs. LPI. An epidermal cell expands enormously in area but only slightly in its height which is just the reversed polarity of growth if compared with the palisade parenchyma.

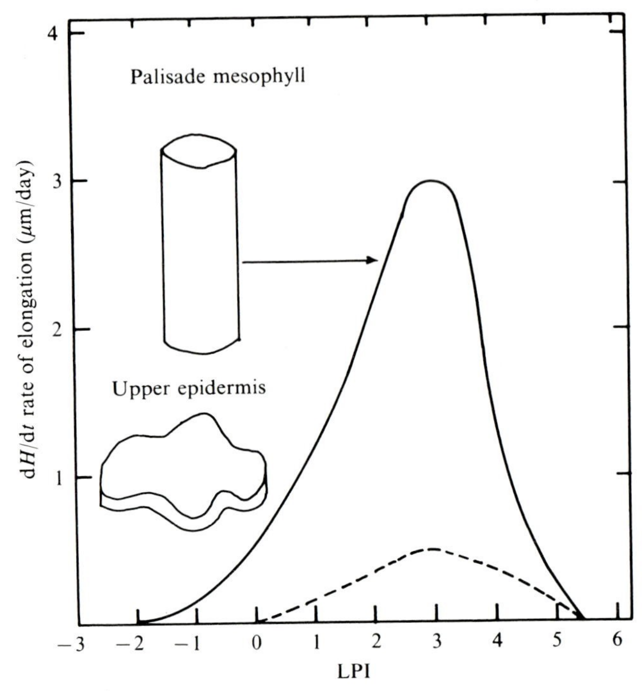

Fig. 36. Rates of elongation of the palisade and upper epidermis plotted vs. LPI. Growth in height of these cells is negligible at early stages of development. Maximum growth is reached at LPI 3.0.

blade in area, and since this is brought about by cell division, and more especially by cell expansion, one would have to approach the whole problem from a cellular point of view in order to get a deeper and more detailed understanding of leaf growth. It is likewise helpful to compare the upper epidermis with the palisade parenchyma and to see what effect one tissue may exert upon another during the process of enlargement and differentiation.

Fig. 37 compares the expansion of cell area of the upper epidermis and palisade mesophyll cells, as seen in paradermal sections. During the embryonic stage the average cell area of the two basic tissues is about the same. This is in agreement with leaf anatomy at the early stages of lamina development, which reveals that cells of the two tissues are pressed against each other in distinct compact layers. After LPI zero the epidermal cells enter into a phase of rapid expansion in area, whereas palisade cells show a decline in cross section area (Figs. 34 and 37), indirectly suggesting that they divide anticlinally at a higher rate and for a longer period of time. Both tissues stop expanding around LPI 5.0. There is a striking difference at this stage in the section areas of the two types of cells, which on the average is 940 μm^2 for the epidermis and only 154 μm^2 for the palisade.

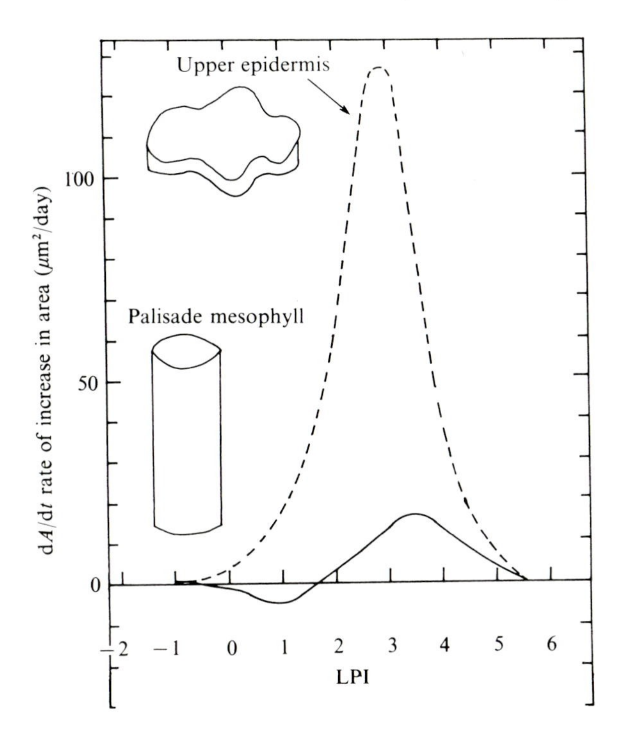

Fig. 37. Rates of increase in cell area of the upper epidermis and palisade cells plotted vs. LPI. Negative rates of expansion at LPI 1.0 are related to the increased rate of anticlinal cell division in the palisade layer. Maximum rate of expansion in cell area is reached around LPI 3.0.

Between LPIs −3.0 and −1.0 the absolute rate increases for the upper epidermis, reaching a maximum of about 125 μm^2 per day at LPI 3.0. With increasing plastochron age the absolute rate decreases sharply and the expansion ceases around LPI 5.0. The palisade cells show a different trend and magnitude from their neighboring tissue. Around LPI 1.0 the absolute rate is negative, indicating that the diameter of these cells has decreased, possibly due to the increasing rate of anticlinal cell division. The maximum absolute rate is much lower here than in the upper epidermis, only 16 μm^2 per day. It is interesting to observe that palisade cells commence their expansion in area about two plastochrons later than the upper epidermis. The expansion of the epidermal cells is of longer duration if compared with the palisade tissue, and it also proceeds at a much higher rate. These two factors, i.e. the duration of the expansion period and the differential rates of expansion in area, may be the primary processes involved in the separation of the palisade mesophyll and the formation of the intercellular spaces.

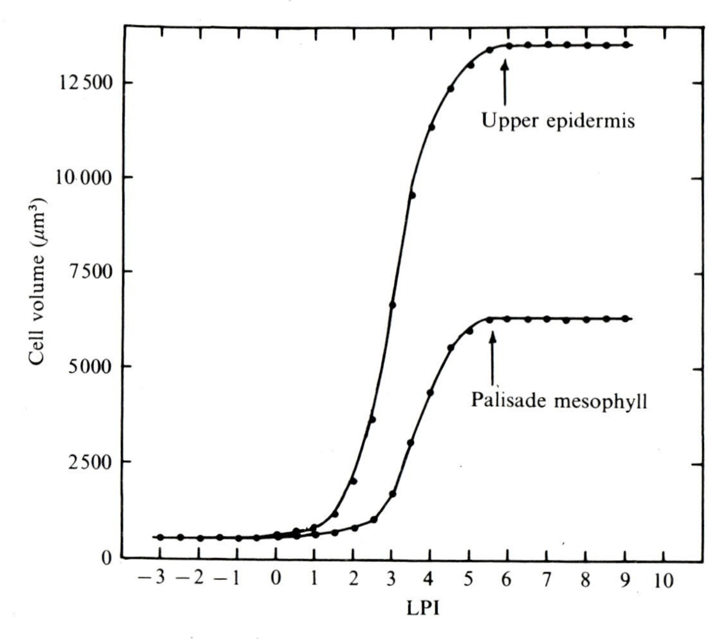

Fig. 38. Cell growth represented by cell volume plotted as a function of the plastochron age. The average volume of the upper epidermal cells and palisade cells in meristematic condition is 460 μm^3 and 576 μm^3; for mature cells it is 13 348 μm^3 and 6729 μm^3, respectively.

In the early plastochrons, there is little difference in the increase in cell volume (Fig. 38) between the meristematic cell of the future palisade mesophyll and the epidermis, each having an average value of 576 μm^3 and 460 μm^3, respectively. The small discrepancy is probably due to the greater height of the palisades. The average nuclear volume of a meristematic cell is about 134 μm^3 which is approximately one-fourth of the average cell volume. After LPI zero, the epidermal cells enter a period of rapid increase in volume, but the palisade parenchyma lags by at least 1.5 plastochrons, commencing its enlargement only after LPI 1.5. The growth in cell volume stops before LPI 6.0 for both tissues. The average mature epidermal cell is about twice as large as its neighboring palisade, the former being 13 348 μm^3 and the latter 6729 μm^3 in volume. It should be realized that the growth of a cell in volume gives only an overall effect. The effect of a two-dimensional expansion is masked here, and only limited conclusions can be drawn on the basis of cellular growth in volume.

Sunderland (1960) has published some data on the average cell volume of sunflower leaves. The shape of the curve of the second pair of leaves is sigmoid and roughly similar to the one in *Xanthium*, even though the values are not directly comparable. The average volume of a meristematic lupine cell is about 940 μm^3 6 days after sowing, and about 1300 μm^3 for a sunflower leaf, 9 days after sowing. Sunderland's values for both meristematic and mature cells are higher than those of *Xanthium* cells. The difference may be due partially to the diverse types of tissues under investigation, and also to differences in environmental conditions. In transition from the meristematic state to maturity, about a 10-fold increase is given for the lupine leaf. For the second pair and the tenth sunflower leaf the increases are 20-fold and 7-fold, respectively. The change from the meristematic to fully differentiated state in the upper epidermis of cocklebur involves a 29-fold, and in palisade, an 11-fold increase in volume.

That the epidermal cell enlargement in the tobacco leaf continues longer than that of any other tissue was reported by Avery in 1933. MacDaniels and Cowart (1944) observed that in an embryonic leaf the ratio of epidermal and palisade cells was 1:1, but that there were 8–10 palisade cells to one epidermal cell in the mature leaf. The unequal growth in the various layers of the lamina (Esau, 1965) is due to different durations of cell division and cell enlargement in the epidermis and other mesophyll layers. From data obtained on *Xanthium* it is possible to expand and modify the above general conclusions in more precise terms and to analyze carefully the course of enlargement of both types of cells.

The palisade cells in embryonic condition are tightly packed polygonal cells approximately 10 μm high and 8 μm diameter. At LPI zero these cells have already commenced their growth in height and can be distinguished in the embryonic leaf. Not only is expansion in area delayed for at least two plastochrons but there is a marked decrease in diameter of the palisade cells which may be attributed indirectly to prolonged cell division and an increase in rate of cell division during this developmental period. It can be inferred that those cells which are dividing are also elongating in height. They also will increase in volume during this critical two-plastochron period. On an absolute rate basis, there is more growth in height than in diameter but the palisade cell stops its expansion in both planes at the same time. It should be stressed that the absolute and relative rates of expansion of both types of cells are not constant or uniform throughout the entire period of development. Each curve has an acceleration phase, a maximum and a deceleration period of growth.

The embryonic epidermal cell is about 9 μm high and 50 μm^2 in area. Its enlargement in both planes starts around zero LPI and at maturity it has grown to be 14 μm high with an area of 940 μm^2. Considering growth on a comparative basis, the palisade cell commences to elongate in height at least one plastochron sooner and at a much higher rate than the epidermis. The upper epidermis has a longer period of expansion in area, and what may be of great significance is that epidermal cells enlarge at higher absolute and relative rates than the palisade mesophyll. Differential rates of expansion, in addition to different duration of cell division and cell enlargement may be the primary factor responsible for the separation of mesophyll cells and the formation of intercellular spaces and tissue differentiation.

12

Correlation of developmental processes

To obtain a more co-ordinated picture of morphological and cellular development, relative rates of cell formation, increase in area, leaf elongation, and the relative rate of expansion of the lamina in thickness are plotted in Fig. 39 against leaf plastochron index. It is convenient to use relative rates for the purpose of correlation of developmental activities since all data can be plotted on the same co-ordinate system. In addition, the relative rate multiplied by 100 is indicative of percent of growth per unit time.

The relative rates of cell formation and increase in area, around LPI 1.0, reach a value of about 34 and 39% of growth, respectively. Leaf elongation at the same plastochron is about 24% per day. Between plastochrons −4.0 and zero the relative rate of expansion of the lamina in thickness indicates that at this period of time cells of the lamina elongate very little. This is indeed in agreement with cytological observations, since at this stage cells are highly meristematic and the absolute rate of growth of individual palisade and epidermal cells is also negligible. The absolute rate of increase in thickness is only 1 μm per day. At LPI 2.0 the maximum relative rate is about 14% per day. The analysis of the expansion of the lamina in thickness represents one dimension of growth. The leaf length and the increase in area are the other variables. Some error is involved in estimating the exact cessation of different growth processes because the data were smoothed by a numerical procedure and all rates were also derived by numerical differentiation. However reasonable approximations can still be quoted. All cell divisions stop shortly after LPI 3.0. The mature thickness of the lamina is established around LPI 4.5. Around LPI 6.0 the blade stops expanding in surface and the elongation of the lamina also ceases. Finally, between LPIs 7 and 8, the petiole stops elongating and at that plastochron age the leaf definitely has reached its maturity.

From correlation studies of morphological development, it becomes evident that leaf development represents a highly organized system. The rates of developmental processes are regulated during the course of development, while a precise sequence of cessation of developmental activities seems to respond to some built-in control mechanisms. It is also evident that physiological and biochemical processes are precisely integrated into morphological development. Khudairi and Hamner (1954), for instance,

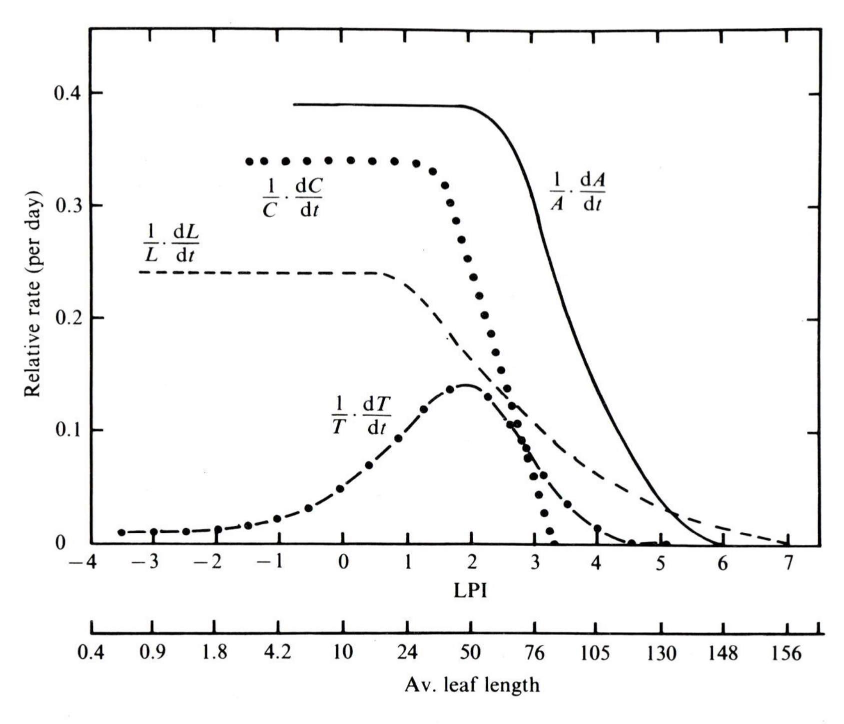

Fig. 39. In this composite graph, the relative rates of cell formation, expansion in area, leaf length and thickness of lamina are presented as a function of the plastochron age (LPI). The average leaf length on the bottom is indicated for corresponding LPIs. Cell formation, increase in area and leaf length at early plastochrons reach values of about 34, 39 and 24% per day respectively. Increase in the lamina thickness is negligible. Cell division stops first, followed by increase in lamina thickness, leaf area and finally leaf length.

investigated sensitivity of *Xanthium* leaves of various stages to photoperiodic induction. They demonstrated that the expanding leaf of an area of 10–15 cm^2 gave the highest flowering response. Salisbury (1955) also indicated that the most rapidly growing half-expanded leaf, between 5.9 to 9.2 cm in length, was maximally sensitive to photoperiodic induction. The leaf ages designated by the area and leaf length can be recalculated into the plastochron ages and correlated with the developmental condition. It appears that *Xanthium* leaves are most photoperiodically responsive around LPI 3.0, which is the stage of rapid elongation of the leaf and most rapid cellular differentiation of the lamina. Studies by Watson (1948) indicate a strong correlation between sensitivity to herbicidal activity and leaf development. The most severe injury from 2,4-D in bean leaves occurs at the time of cell differentiation of the laminar tissue.

II

PHYSIOLOGICAL ASPECTS OF DEVELOPMENT

13

Introduction

Up to this point, the morphological aspects of development have been considered. The plastochron index was derived to provide a precise yet practical scale to stage the various developmental parameters. Organization of the shoot apex, leaf initiation, the origin and development of the lamina, the thickness of the lamina and leaf area were considered on a quantitative basis to give a precise description of leaf growth. Cell division, cell lineage in the marginal and plate meristems, together with cell enlargement and tissue differentiation, constituted the cellular aspects of development. Finally, the correlation of developmental processes was intended to put the morphological and cellular aspects of development on a unified basis. An attempt was made not only to specify 'when' during development but also 'how much' and 'at what rate'.

The complexity of leaf development increases with the introduction of physiological and biochemical aspects of development. These will be dealt with in the subsequent chapters. Since a substantial amount of experimental work was done with labeled nucleic acid precursors and autoradiographic techniques, we felt that it may be appropriate, even though a little unconventional, to discuss the techniques and the basic kinetics of ^{3}H-thymidine incorporation, and thus to provide a basis for elucidation of other developmental processes, such as DNA biosynthesis. Several arguments prompted us to do so. Relatively little work has been published on the kinetics and procedures of incorporation of labeled DNA precursors with special emphasis upon leaf tissues, and leaf development in particular. In addition, this discussion will provide the readers with the necessary background for the discussion of the synthesis of nuclear DNA in various leaf tissues. Chloroplast growth, chlorophyll synthesis and respiration will be discussed in the last chapters under consideration. However, an attempt will be made to correlate these processes to such developmental processes as cell differentiation, enzymatic activity, protein synthesis and senescence.

14

Incorporation of ^{3}H-thymidine into nuclear DNA

Tritium-labeled thymidine has been used extensively to investigate cell division in a variety of cells and tissues. It has been established that thymidine is incorporated into nuclei of cells synthesizing DNA prior to mitosis. Although the literature on ^{3}H-thymidine application is quite extensive, little quantitative information is available on its uptake by nuclei of leaf cells or on the methods of its incorporation. To achieve a high degree of reproducibility from one experiment to another, this information is essential. In the labeling procedures several variables must be considered: the concentration of the radioisotope, its specific activity, the incubation time in the radioisotopic solution and the time of growth after treatment. It is essential, therefore, to establish the basic kinetics of incorporation of the tritium-labeled thymidine into the nuclei of cells of young leaves before considering the pattern of incorporation and DNA biosynthesis in various stages of development.

Since detailed autoradiographic methods have been described elsewhere, (Jensen, 1962; Maksymowych and Blum, 1966*a, b*; Rogers, 1967; Maksymowych, Devlin, Blum and Wochok, 1967) only the essential procedures applicable to leaves will be outlined.

Xanthium leaves between LPIs -0.5 and 0.5 should be selected, since at this stage the leaves are characterized by six compact meristematic layers of cells and active cell division. Several methods of application of the ^{3}H-thymidine were tried, but most were not successful. The best results have been obtained by foliar absorption of the radioisotope. In this method, the whole shoot with young leaves or one desired leaf is submerged in 0.01% 'Tween' or 'Tergitol' for about 15–20 minutes to facilitate uptake of the isotope (Juniper, 1959). The wetting agent is replaced by ^{3}H-thymidine with a concentration of 10 μCi/ml for about 2–4 hours. Air is bubbled into the medium to increase oxygen concentration. After the period of exposure to the radioisotopic solution, the leaf is rinsed with distilled water and, if desired, the plants can be grown for some time in the absence of ^{3}H-thymidine to allow time for translocation and incorporation. After fixation and infiltration with Paraplast, the tissue is cut into 4 μm thick sections and stained with Feulgen. Kodak stripping film AR-10 or Kodak Nuclear Track

Emulsion NTB-2 can be used in autoradiography. In general, better results were obtained with the liquid emulsion. Slides are exposed usually for two weeks at 4°C and processed according to established autoradiographic techniques (Jensen, 1962; Rogers, 1967). A Bausch and Lomb Whipple micrometer can be used for grain counts or estimation of the percent of labeled nuclei. Recently Rogers (1967) and Dormer, Brinkmann, Stieber and Stich (1966) developed a cytospectrophotometric method of the assessment of the number of grains. In this technique, dark-field illumination of the autoradiograph converts the silver grains into bright specks. The light reflected back towards the objective by the silver grains is collected and directed onto the photocathode of a photomultiplier tube. The current generated by the photomultiplier is recorded on a galvanometer. The number of silver grains per unit area is equated as a linear function with the photometer values.

In quantitative autoradiography perhaps the most useful approach in collecting data is the assessment of the average number of grains per nucleus and the estimation of the average percent of nuclei labeled. Both of these variables indicate a relative amount of the radioactive precursor incorporated into DNA molecules. The percent of nuclei labeled with ^{3}H-thymidine will represent the relative number of cells in a cell population engaged in DNA synthesis. As evident from Table 4, the percent of labeled nuclei was derived from a cell population which included both labeled and unlabeled nuclei. The number of grains per nucleus, however, will be indicative of the relative amount of the radioactive precursor incorporated into the nuclear DNA of a single cell. Autoradiographs of small parts of cross sections of laminae are illustrated in Figs. 40, 41 and 42.

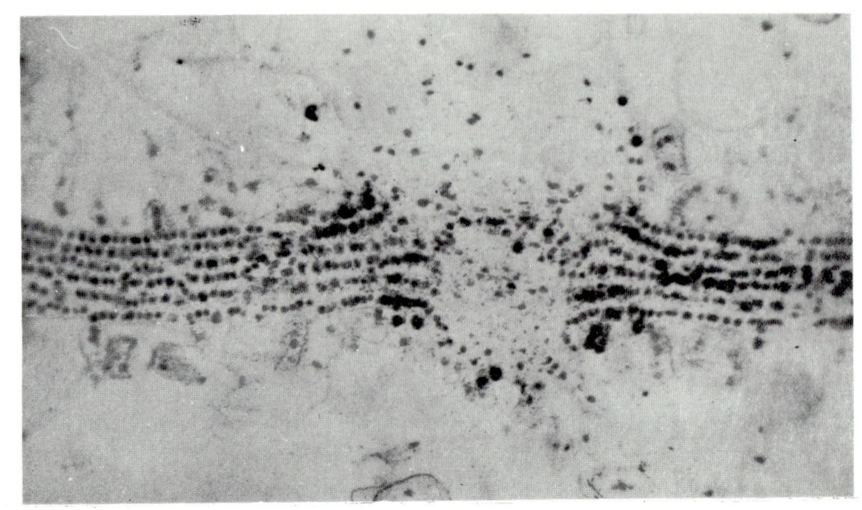

Fig. 40

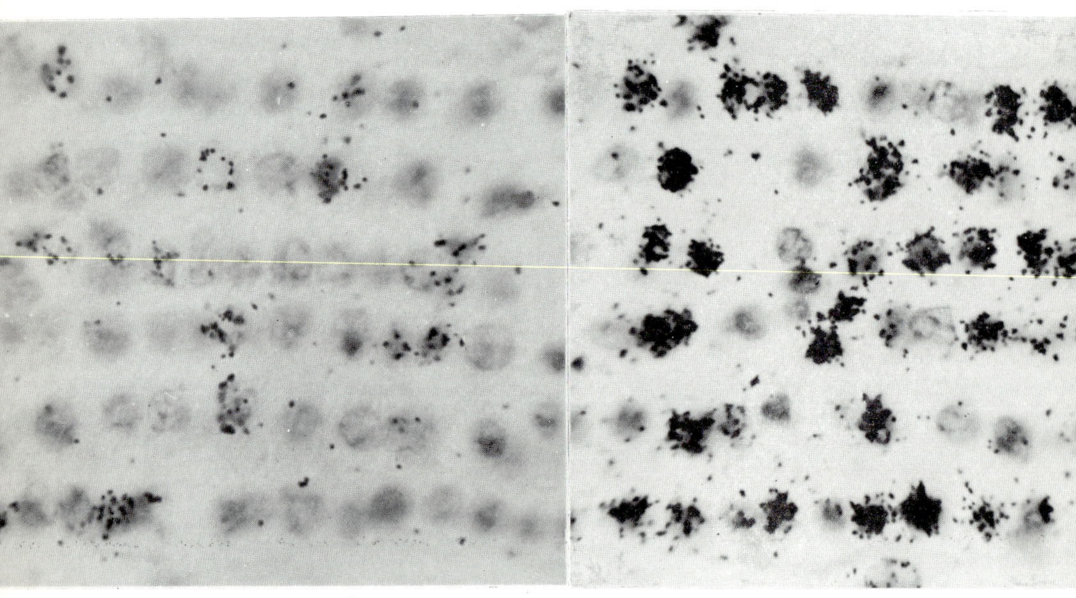

Fig. 41 Fig. 42

Figs. 40–42. Photomicrographs of *Xanthium* leaves with ^{3}H-thymidine labeled nuclei.

Fig. 40 is a low-power photomicrograph of a young leaf showing heavier concentration of label near a prominent vein. Fig. 41 is a high-power photomicrograph of a young leaf. Average diameter of a nucleus is about 6.2 μm. The grains over nuclei represent relative amounts of tritiated thymidine incorporated into nuclear DNA. The leaf was treated for four hours with 10 μCi/ml of the radioisotope with specific activity 6.7 Ci/mmole. Fig. 42 shows heavier label in the leaf which was incubated for 16 hours in the radioisotopic solution.

15

Kinetics of ^3H-thymidine incorporation

One of the serious questions in developmental studies is the possibility that some deleterious effects of beta radiation might disturb the normal course of development. No detectable changes on the gross morphological level were observed in relation to the leaf length and morphology of the plants treated with ^3H-thymidine, thymidine and control. Neither the plastochron index nor the leaf length of the experimental plants (Fig. 43) was affected by various treatments. However, on the cytological level (Fig. 44) the mitotic index dropped significantly if plants were grown in the radioisotope for more than eight hours. The average 8% of mitosis obtained in controls decreased to less than 5% in experiments with more than eight hours of growth in the radioisotopic solution. Prolonged incubation of leaves in 10 μCi/ml of ^3H-thymidine was manifested in the decline of cell division, indicating probable interference with the mitotic apparatus.

The average number of grains per nucleus increased as a linear function of concentration of the radioisotopic solution (Fig. 45) expressed in μCi/ml. Concentration of 10 μCi/ml and four hours of growth in the radioisotopic solution showed that, on the average, about 20% of the nuclei were labeled and that there were 18 grains per average labeled nucleus. These are suitable magnitudes for visual counts and statistical computations. Nuclei with five grains and more were usually considered as specifically labeled. A seven-fold increase in concentration of the isotope gave a two-fold increase in the average number of grains per labeled nucleus. It is possible that, with an influx of the exogenous ^3H-thymidine, some of it is pumped into the vascular system and is not immediately available for the incorporation. A physiologically active plant organ in this case may deviate significantly from the theoretically expected stoichiometric relationship. Another interesting observation based on Fig. 45 is the fact that there was no significant difference among the various specific activities. This leads to the conclusion that various specific activities of ^3H-thymidine (2, 6.7 and 13.7 Ci/mmole) had no effect upon the amount of the radioisotope incorporated, provided the amount of μCi/ml was kept constant. For a more detailed explanation of the effect of specific activity, the reader is referred to Maksymowych *et al.* (1966*a*, *b* and 1967).

The average number of grains per nucleus (Fig. 46) increased logarith-

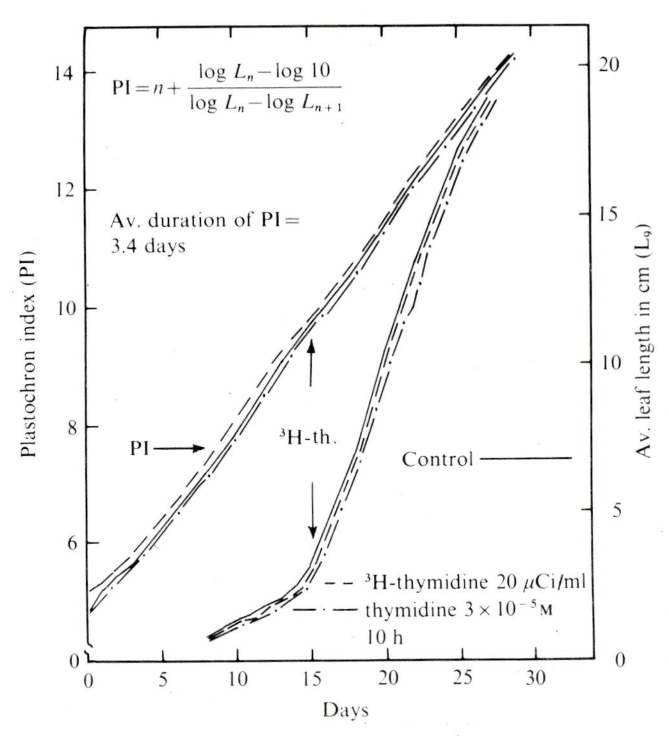

Fig. 43. The effect of ^{3}H-thymidine upon leaf elongation. Plastochron index and the average leaf length remained unchanged in ^{3}H-thymidine and thymidine treated plants as compared with the control.

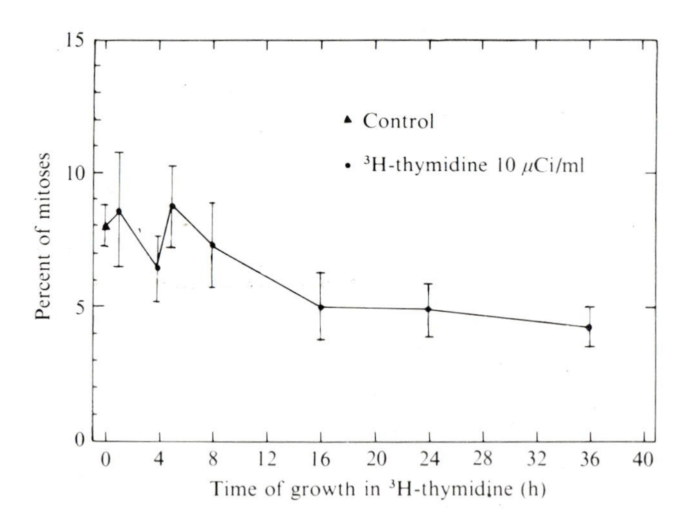

Fig. 44. Percent of cell division plotted vs. the time of exposure to ^{3}H-thymidine indicates no significant differences between treated and untreated leaves. However, the mitotic index dropped significantly if plants were grown in the isotope for more than 8 hours.

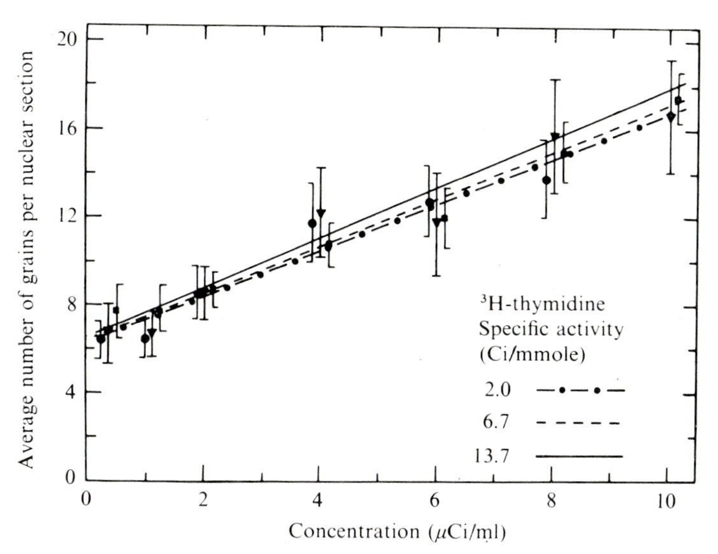

Fig. 45. Average number of grains per labeled nuclear section is presented as a function of ^{3}H-thymidine concentration with various specific activities. A seven-fold increase in concentration gave a two-fold increase in grain counts. Nuclei with six grains or more were considered as specifically labeled. Slopes of the fitted lines would change if different criteria of grain assessment were used.

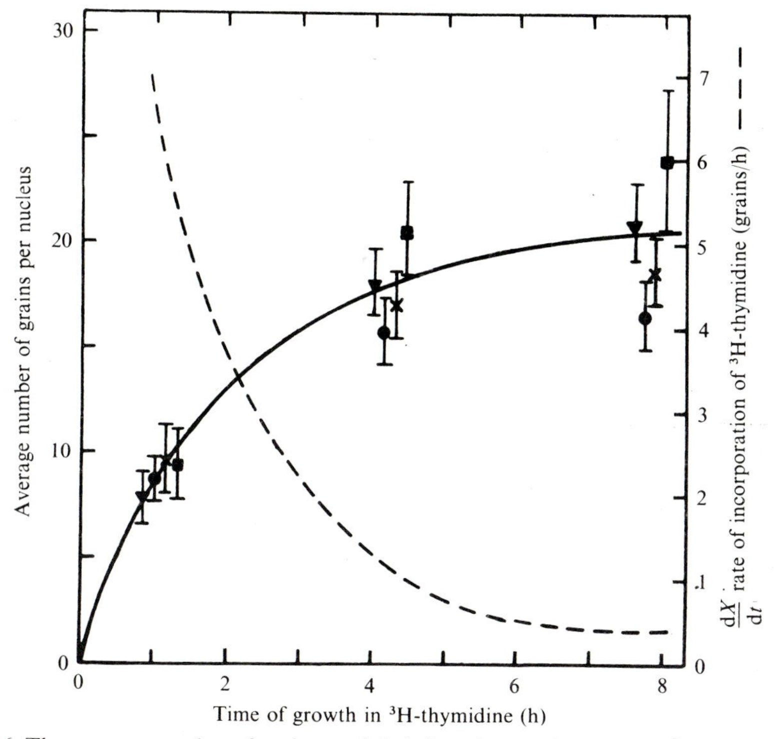

Fig. 46. The average number of grains per labeled nucleus and the rate of ^{3}H-thymidine incorporation plotted vs. time of exposure to the radioisotope. The solid line approximates a logarithmic function. Rates of incorporation are high during the first hour, rapidly decelerating thereafter.

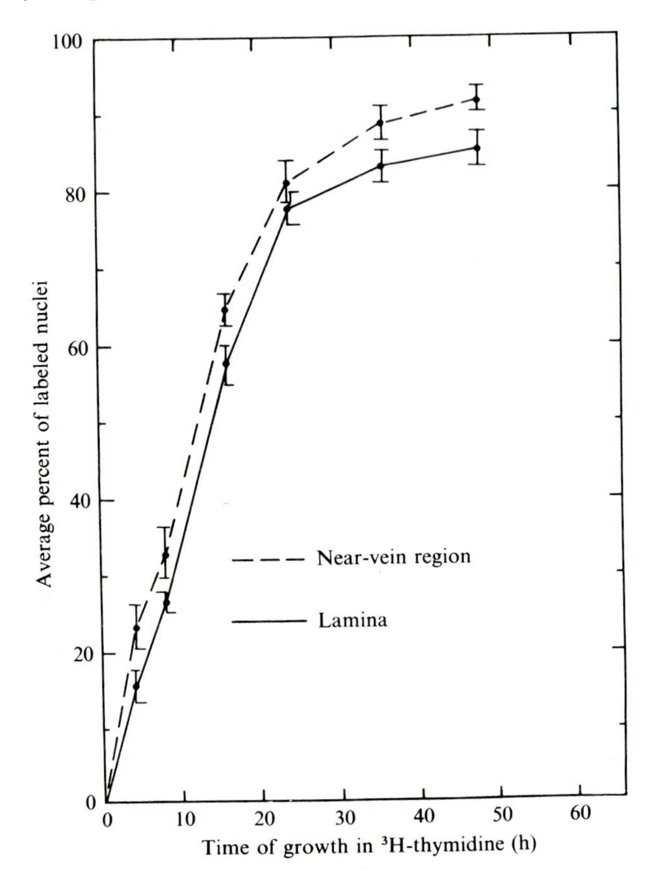

Fig. 47. The average percent of nuclei labeled with ^{3}H-thymidine plotted vs. time of growth in ^{3}H-thymidine, for the lamina region near a prominent vein and for the lamina region at some distance from the vein. The near-vein region incorporated significantly higher amounts of the label.

mically as a function of time of exposure to ^{3}H-thymidine. These data also support the conclusions given in the previous paragraph, that at the specified concentrations, the specific activities had no effect upon the amount of tritiated thymidine incorporated.

The average percent of labeled nuclei, as illustrated in Fig. 47, increased linearly as a function of time of growth in the radioisotope. After twenty-two hours the curves leveled off, showing that over 80% of the nuclei were labeled. Another noteworthy observation is that the lamina is labeled through its entire extent from the midrib to its margin. Quantitatively, the grain density is not uniform in various parts of the lamina. The near-vein region, which is a six-layered part of the lamina about 70 μm long and is just adjacent to a

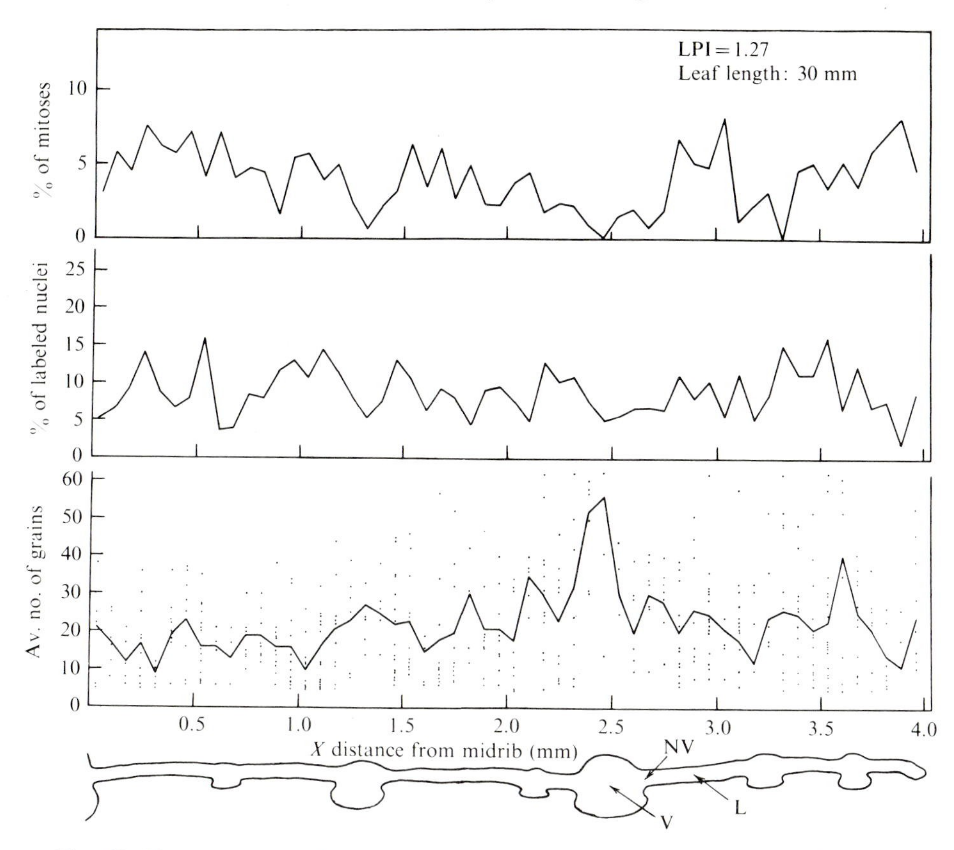

Fig. 48. The average number of grains per labeled nucleus, percent of nuclei labeled with ^{3}H-thymidine, and percent mitosis of a *Xanthium* leaf plotted vs. the distance from the midrib. V – vein; NV – near-vein region of the lamina; L is six-layered lamina. The lamina is labeled throughout its extent; however, there are significant differences of ^{3}H-thymidine incorporation in various parts of the lamina.

major vein, incorporates significantly higher amounts of the label. Any part of the six-layered lamina, which is free of vein and referred to as the lamina region, is labeled less extensively than the near-vein region. Veins, like near-vein regions, incorporate significantly larger amounts of the radioactive DNA precursor per average labeled nucleus than the lamina regions.

Higher incorporation of the exogenous thymidine in the near-vein region may be attributed to a higher rate of nucleic acid metabolism in this region. This in turn might indicate that the rates of lamina extension on a micro-level are not evenly distributed throughout the lamina and that the near-vein region is a specialized growth center. It is also possible that the demonstrated differences in labeling are due to the availability of the exogenous precursor supplied by the vascular system during the period of growth in the radio-isotope. In this case, the near-vein region would receive more of the exogenous precursor simply because its cells are located in the proximity of the vein.

Labeling experiments in which the entire lamina was labeled (Fig. 48) point toward some practical application in the assessment of the autoradiographic data obtained on the various parts of the lamina. As evident from the distribution of label, random samples can be taken at any distance from the midrib. However, some complications are encountered at the vein and near-vein regions. There are two possibilities of data assessment. All nuclei could be scored through the extent of the lamina, or data sampling could be limited only to six-layered regions of the lamina, excluding vein and near-vein regions. Since the first method is excessively laborious, the second approach appears to be more practical and can be used with greater efficiency.

16

Synthesis of nuclear DNA in various tissues

The autoradiographic techniques can be successfully applied in studies of nucleic acid synthesis during leaf development. They are especially suitable for following DNA biosynthesis localized in the epidermal, palisade and spongy mesophyll layers of cells. These laminar tissues cannot be separated easily for chemical extraction by procedures which were developed by Schmidt and Thannhauser (1945) for animal tissues, and subsequently modified by Ogur and Rosen (1950) and Smillie and Krotkow (1960) for plant tissues.

Avery (1933) and MacDaniels and Cowart (1944) reported that during development there are time differences in the duration of cell division in the epidermis and in the various mesophyll layers of tobacco and apple leaves. In addition to the duration of cell division and enlargement, an important factor in shaping final morphology of the *Xanthium* leaf is the rate with which the different cells divide and elongate at a particular developmental stage (Maksymowych, 1963). Since DNA synthesis is an integral part of the mitotic cycle, an attempt was made to describe quantitatively the pattern of DNA synthesis for the upper and lower epidermis, the palisade layer, and the spongy mesophyll cells.

Fig. 49 represents the average percent of labeled nuclei on the left ordinate and the average number of grains per labeled nucleus on the right ordinate, plotted as a function of the plastochron age of the leaf. At LPI -0.56 (Table 4) the two curves reach a value of about 22% of the labeled nuclei and 22 grains per nucleus. The functions decrease somewhat in the course of development, giving at LPI 1.72, on the average, 16% of the labeled nuclei and 15 grains per nucleus. One may conclude that the amount of DNA synthesis decreases with the age of the leaf in the course of development. This means that at LPI 1.7 fewer cells are engaged in DNA synthesis and also that a smaller amount of the radioactive precursor is incorporated in a single labeled nucleus than at LPI -0.56. After LPI 1.7 the amount of DNA synthesis decreases very rapidly. In leaves older than LPI 2.5, only a small number of cells are engaged in nucleic acid synthesis. However, the amount of incorporation of the radioisotope on a single cell basis increases very rapidly starting with LPI 1.7. It can be concluded that at LPI 2.5 fewer cells

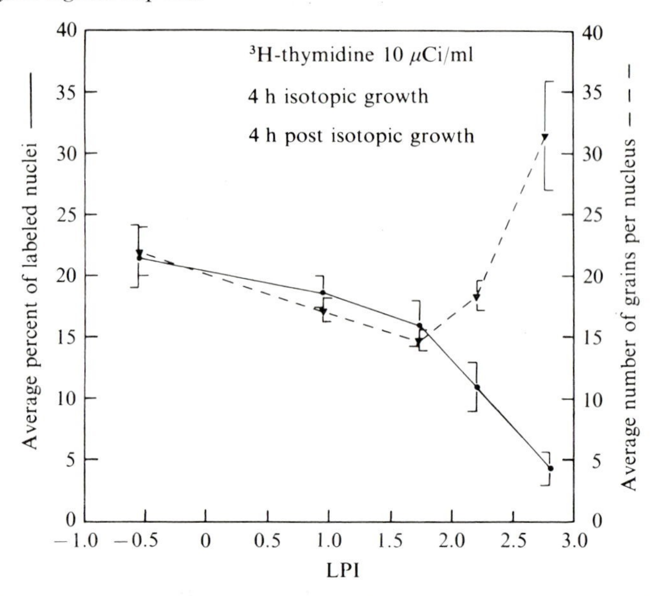

Fig. 49. The average percent of nuclei labeled with ^{3}H-thymidine and the average number of grains per labeled nucleus are plotted as functions of the leaf plastochron index. They represent a developmental pattern of DNA synthesis and ^{3}H-thymidine incorporation in the leaf lamina.

are engaged in DNA synthesis, but they incorporate more ^{3}H-thymidine into DNA of a single nucleus than at any younger plastochronic age. It should be emphasized at this point that the leaf lamina at LPI −0.5 consists of six compact layers of meristematic cells characterized by active cell division. The cells of the middle portion of the lamina at LPI 1.7 are considerably enlarged and many of them in a state of differentiation. The rate of cell division at this plastochron is significantly smaller than in a younger lamina.

Up to now DNA synthesis has been analyzed in a cell population as seen in a cross section of the entire lamina. No reference has been made to the individual tissues. The middle portion of a young lamina comprises the upper and lower protoderm layers which are future epidermis, adaxial and abaxial layers, and two middle layers. The adaxial layer will give rise to the future palisade mesophyll, and the abaxial and middle layers will differentiate into the spongy mesophyll tissue and vascular tissue. The grains over some nuclei in Fig. 50 represent the labeled nuclei which have incorporated ^{3}H-thymidine into DNA molecules. Each layer of cells has at least one labeled nucleus. At LPI −0.56, the ratio of nuclei in the adaxial and upper protoderm layers is about 1:1.

A more advanced stage of development at LPI 1.93 is pictured in Fig. 51. The upper epidermis at this stage is not labeled, indicating that no DNA

Table 4. Incorporation of ^{3}H-thymidine into nuclear DNA of the entire lamina

Average LPI	Number of leaves analyzed	Total number of analyses	Total number of nuclei scored	Number of nuclei with 6 grains or more	Percent of labeled nuclei	95% Confidence intervals (C.I.)	Average number of grains per nucleus	C.I.
−0.56	3	24	1269	273	21.6	±2.6	22.0	±2.0
+0.94	5	36	2995	557	18.6	±1.3	17.2	±0.9
+1.72	5	40	7713	754	16.0	±1.9	14.8	±0.8
+2.20	4	40	7750	485	10.9	±2.1	18.5	±1.2
+2.78	3	24	2190	92	4.2	±1.4	31.6	±4.5

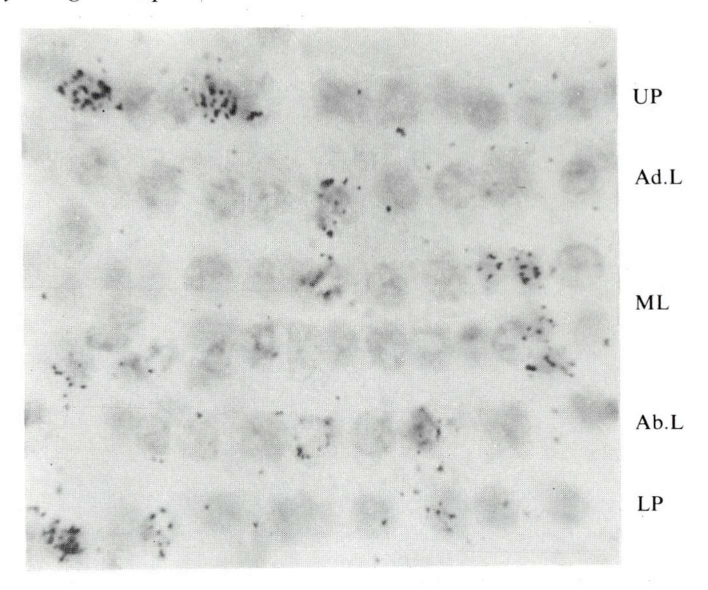

Fig. 50. Autoradiograph of a cross section of *Xanthium* leaf at LPI −0.56 which was treated for 4 hours with ^{3}H-thymidine. The middle portion of a young lamina comprises the upper (UP) and lower (LP) protoderms, the adaxial (Ad.L) and abaxial (Ab.L) layers and the two middle layers (ML). The grains over some nuclei represent the labeled nuclei which are engaged in DNA synthesis. All layers are labeled. The frequency of nuclei in the upper protoderm and adaxial layers, per unit of leaf area is about the same.

synthesis takes place in this cell layer. Many palisade cells however, are heavily labeled. The relative frequency of the nuclei per unit leaf area in the upper epidermis and palisade layer indicates that there are more than twice as many palisade cells as upper epidermal cells. This may indicate that the prepalisade layers went through one or two cycles of division more than the protoderm.

Fig. 52 represents the average percent of labeled nuclei of the upper and lower epidermis, the middle mesophyll tissue, and the palisade layer plotted as a function of the plastochron age of the leaf. The amount of DNA synthesis is very similar in the upper and lower epidermis. At LPI −0.56 (Table 5), the curves which represent the epidermal layers start with about 22% of the labeled nuclei. In the course of development this value decreases constantly, showing that at LPI 1.72 an average of about 15% of the nuclei were labeled. It appears that the amount of DNA synthesis in both epidermal layers significantly decreases at this period of development. More epidermal cells are engaged in the nucleic acid synthesis at LPI −0.56 than at LPI 1.72. In older leaves the decline in the percentage of labeled nuclei becomes steeper, approaching zero at about LPI 3.0. The curve representing ^{3}H-thymidine incorporation in the middle mesophyll has a trend similar to the

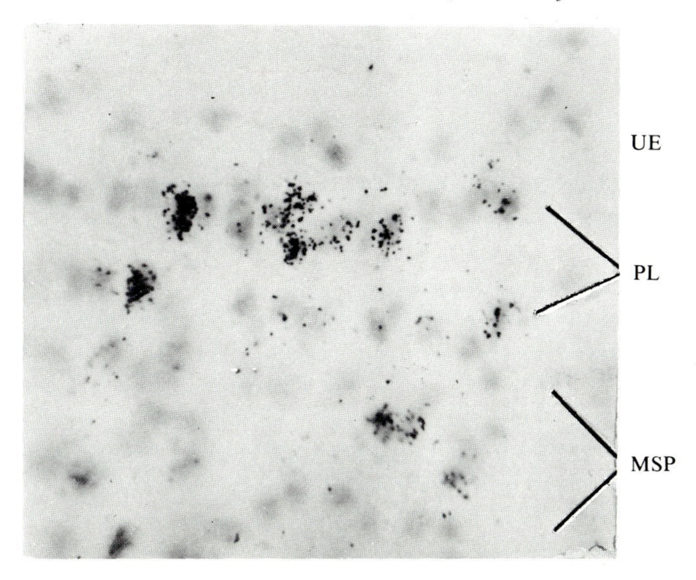

Fig. 51. An autoradiograph of a cross section of leaf lamina representing an advanced stage of development at LPI 1.93. The upper epidermis (UE) at this stage is not labeled indicating that no DNA synthesis takes place in this tissue. A large proportion of the palisade cells (PL) is heavily labeled, and the middle spongy parenchyma (MSP) is lightly labeled. There are approximately twice as many palisade cells per unit of leaf area than upper epidermal cells as seen by the relative frequency of the nuclei.

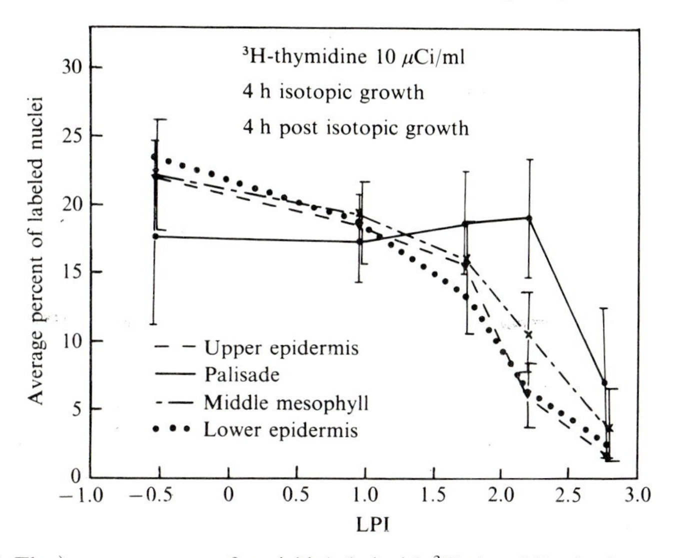

Fig. 52. The average percent of nuclei labeled with ^{3}H-thymidine in the upper and lower epidermis, palisade and mesophyll tissues plotted vs. leaf plastochron index. The palisade layer has a different pattern of DNA synthesis from the other tissues.

Table 5. Incorporation of ^{3}H-thymidine into nuclear DNA of various lamina tissues

Average LPI	Number of leaves	Number of analyses per tissue	Upper epidermis: Total number of nuclei scored	Number of nuclei with 6 grains	Percent of labeled nuclei	95% Confidence intervals (C.I.)	Average number of grains per nucleus	C.I.	Mesophyll: Total number of nuclei scored	Number of nuclei with 6 grains	Percent of labeled nuclei	C.I.	Average number of grains per nucleus	C.I.
−0.56	3	6	308	68	22.1	±4.1	22.5	±4.3	242	43	17.8	±6.9	19.9	±4.4
+0.94	5	9	717	134	18.7	±3.1	17.0	±1.8	740	128	17.3	±3.3	18.3	±2.0
+1.72	5	10	1025	162	15.8	±4.2	14.1	±1.5	1214	227	18.7	±4.2	14.5	±1.5
+2.20	4	16	1049	64	6.1	±2.3	15.9	±2.6	1109	214	19.3	±4.4	18.4	±1.9
+2.78	3	6	529	9	1.7	±2.3	45.1	±19.3	632	43	6.8	±5.4	28.1	±6.7

Average LPI	Number of leaves	Number of analyses per tissue	Palisade: Total number of nuclei scored	Number of nuclei with 6 grains	Percent of labeled nuclei	C.I.	Average number of grains per nucleus	C.I.	Lower epidermis: Total number of nuclei scored	Number of nuclei with 6 grains	Percent of labeled nuclei	C.I.	Average number of grains per nucleus	C.I.
−0.56	3	6	354	79	22.3	±6.4	21.5	±3.4	353	83	23.5	±7.3	23.2	±3.7
+0.94	5	9	826	161	19.5	±2.7	15.6	±1.6	705	134	19.0	±3.5	18.4	±2.0
+1.72	5	10	1363	218	16.0	5.2	14.7	±1.6	1089	147	13.5	±3.0	16.3	±2.1
+2.20	4	16	1271	136	10.7	±2.8	19.2	±2.2	1109	71	6.4	±2.2	20.0	±3.3
[illegible]	[illegible]	[illegible]	[illegible]	[illegible]	[illegible]	[illegible]	[illegible]	[illegible]	[illegible]	[illegible]	[illegible]	[illegible]	[illegible]	[illegible]

epidermal tissue. Only at LPI 2.2 is the amount of incorporation in the mesophyll somewhat higher than in the epidermal layers. DNA synthesis in the palisade layer, however, follows a different pattern in the course of leaf development. Between LPIs −0.5 and 0.94 the amount of incorporation of the radioactive DNA precursor is smaller than in the other tissues. Between LPIs 0.94 and 2.2, an almost linear increase is noticeable. At the latter plastochron age, there is a highly significant difference in the amount of DNA synthesis in the palisade cells compared to cells of the other tissues. Only after LPI 2.2 does the curve drop down very rapidly, approaching zero at about LPI 3.0. The DNA synthesis in the palisade layer between LPIs 1 and 2 is higher than in the upper and lower epidermis and the middle mesophyll. It should be noted that this synthesis stops around LPI 3.0. After this plastochron, no ^{3}H-thymidine is incorporated into nuclear DNA.

A decrease in the average diameter of palisade cells around LPI 2.0 (Fig. 34) can be attributed to an increased rate of cell division. Significantly higher rates of DNA synthesis in the palisade layer at the same plastochron age provide additional support for the above interpretation, indicating also a good correlation between these two developmental processes.

DNA synthesis in the palisade layer continues for at least 3.5 days longer and at a significantly higher rate than in the epidermal and mesophyll cells of the lamina. If a close analogy can be drawn between cell division and deoxyribonucleic acid synthesis, this finding would be in agreement with the established concept (Esau, 1965) that, in many dicotyledonous leaves, cell division usually ceases first in the upper epidermis and continues longer in the palisade tissues.

17

Metabolism of ^{3}H-thymidine during cellular differentiation

Young *Xanthium* leaves between LPI 0.5 and 0.95 were exposed to ^{3}H-thymidine for eight hours and after treatment they were transferred to the environmental chamber for further growth. The treated leaves were fixed approximately at three-day intervals and the autoradiographic slides were exposed two weeks to the NTB-2 nuclear track emulsion. These experiments were designed to study the long range effect of the labeled nucleic acid precursor and its metabolism during cellular differentiation.

As expected, all autoradiographic slides had labeled nuclei indicating that ^{3}H-thymidine was incorporated into nuclear DNA. However, slides with leaf tissues which had two weeks of growth after ^{3}H-thymidine treatment gave the most interesting results. In addition to labeled nuclei, cell walls of many tissues were intensely labeled. This was apparent in epidermis (Fig. 53), palisade and spongy parenchyma, but most strikingly in xylem vessels. As illustrated in Fig. 54, the label was found in the secondary wall depositions.

If meristematic cells incorporate ^{3}H-thymidine in quantities exceeding the requirements of the endogenous pool, the excess of thymidine is metabolized during cellular differentiation, and its labeled fraction is incorporated into the secondary wall. This example clearly indicates the ability of a cell to regulate and adjust its metabolism to the existing needs. One may speculate that with the accumulation of large amounts of the exogenous thymidine, the steady state of cellular metabolism was altered. The synthesis of the endogenous thymidine was affected and large quantities of reserve thymidine accumulated. Part of the accumulated thymidine was used in the synthesis of nuclear DNA. The rest of it was metabolized, degraded to an undetermined labeled fraction and incorporated into the secondary wall of differentiating cells.

It is interesting to point out that the cell wall label can be detected only in tissues which were exposed to ^{3}H-thymidine in its meristematic condition. Differentiating cells, if exposed to tritiated thymidine, show only a slight cytoplasmic label but no nuclear label and no cell wall label.

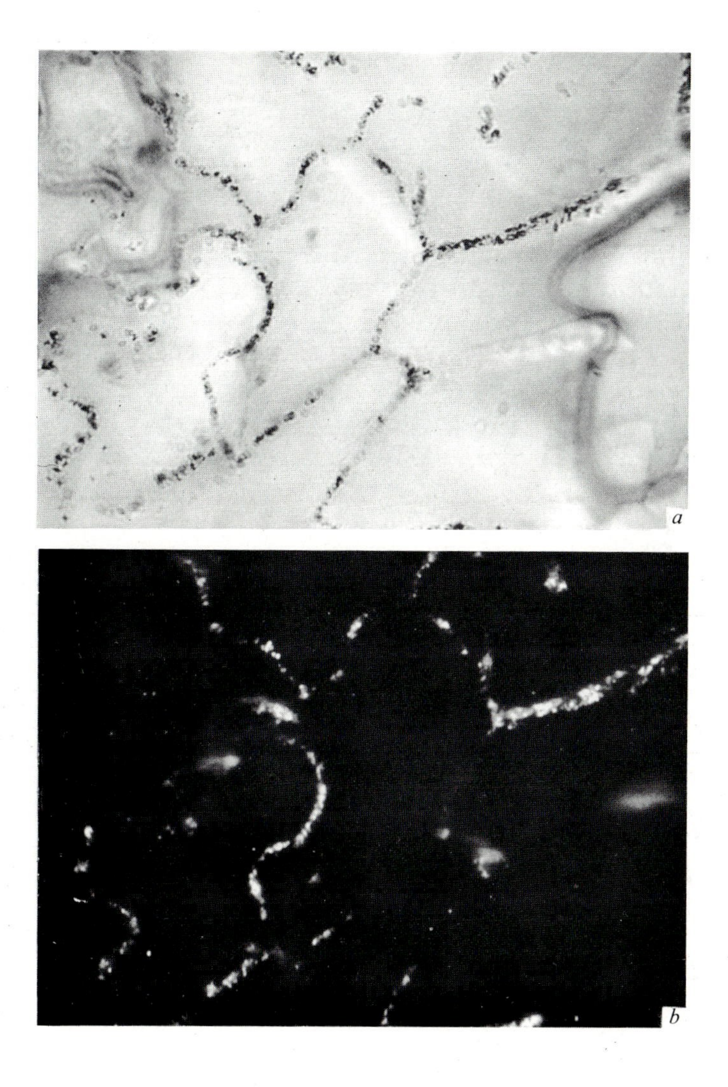

Fig. 53. Autoradiographed paradermal section (*a*) of a *Xanthium* leaf showing labeled cell wall of the upper epidermis. Young leaves were exposed to ^{3}H-thymidine for 8 hours and were given two weeks of growth after treatment. Thymidine was metabolized during cellular differentiation and its labeled fraction was incorporated into the secondary wall depositions. The same leaf section (*b*) was photographed with incident dark-field illumination (810 ×).

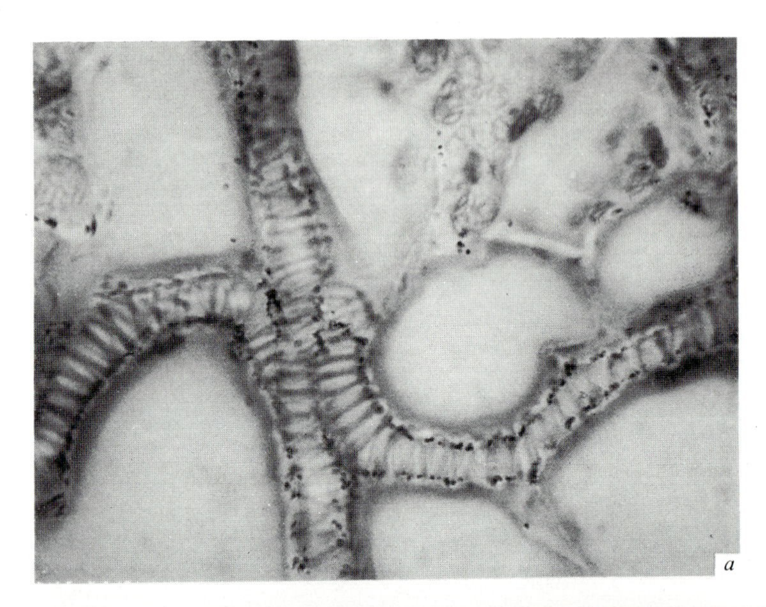

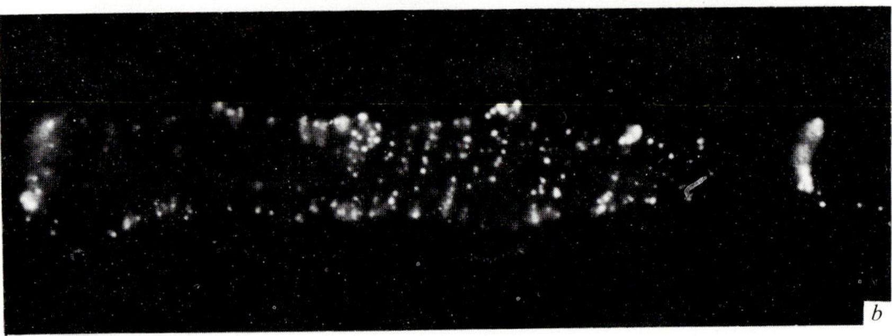

Fig. 54. Autoradiographed paradermal section (*a*) of a *Xanthium* leaf showing label in the secondary wall of a xylem vessel. Treatment was similar as described in Fig. 53. Labeled fraction of thymidine molecule was incorporated into the secondary wall depositions (810×). Autoradiograph of a different xylem vessel (*b*) was photographed with incident dark-field illumination. Spiral thickenings in this vessel are labeled.

18

Chloroplast growth

The relationship between the chloroplast size and leaf development was studied by Holowinsky *et al.* (1965) in *Xanthium* and red kidney bean. With the fluorescence microscope they observed that cells of youngest leaf primordia, at LPI -6.0, contained numerous red fluorescing particles of a diameter of about 1.6 μm as illustrated in Fig. 55, the progressively older primordia contained larger particles. With the increasing age of the leaf, the diameter of these particles increased and at maturity reached an average value up to about 6 μm. A change in the shape of the plastids from circular to ellipsoid was concomitant with their growth.

The rate of increase in an average chloroplast diameter in relation to the increasing morphological age of the leaf is low in the range of LPI -6.0 to LPI 2.0. Assuming a linear relationship, the rate at this plastochron range is about 0.04 μm per day. The rate of increase between LPIs 2.0 and 6.0 is almost six times as high or about 0.25 μm per day. The rapid increase in chloroplast growth occurs between LPIs 0.0 and 2.0. About 70% increase in the chloroplast diameter takes place above LPI 2.0.

The major change of chloroplast growth takes place after most cell divisions in the leaf have been accomplished, while the highest rate of plastid growth takes over during the period of cellular differentiation. It is perhaps also of significance that the rapid chloroplast growth coincides with the maximum rate of increase in leaf area around LPI 3.5 and the maximum rate of cell growth at about LPI 3.0. The pattern of chloroplast growth parallels closely the growth of the lamina and in particular the cell enlargement. One is led to conclude that chloroplast growth is in some way closely tied to cell enlargement and lamina expansion. It also appears that the regulatory aspects of this subcellular organelle development is closely integrated with cellular and organ development.

19

Chlorophyll synthesis

The pattern of chlorophyll synthesis in the *Xanthium* leaf parallels closely the increase in surface area. A semilogarithmic plot of the amount of chlorophyll per leaf (Fig. 56) gives a straight line in the early plastochrons and levels off after LPI 4.0, during the subsequent development. This indicates that chlorophyll synthesis follows an exponential increase in young developing leaves, achieving a steady state at maturity. A young leaf of LPI 0.0 which is 10 mm long, as estimated by Michelini (1958), contains about 7 μg of chlorophyll. Since the average fresh weight of leaf 9 at this plastochron is about 11 μg, Michelini's data would yield about 0.64 μg of chlorophyll per mg fresh weight. Holowinsky *et al.* (1965) have observed a sharp increase in chlorophyll synthesis (Fig. 55) above LPI 2.0. Both authors

Table 6. Chlorophyll in *Xanthium* leaf

1	2[a]	3	4[b]	5	6
LPI	Av. number of cells in leaf (× 10⁴)	Number of photosynthetic cells in lamina (× 10⁴)	Chlorophyll		Rate of change in chlorophyll synthesis (pg/cell/day)
			μg/leaf	pg/cell	
−2.5	8	4.7			
−2.0	16	9.4	0.17	1.88	
−1.5	35	20.6	0.45	2.18	0.55
−1.0	66	32.5	1.15	3.53	0.66
−0.5	135	69.9	2.91	4.16	0.58
0.0	262	140.1	7.4	5.28	0.61
+0.5	560	307.7	18.4	5.98	0.81
1.0	1100	617.5	47.6	7.71	1.10
1.5	2300	1307.4	121.3	9.28	1.40
2.0	4500	2578.3	306.9	11.90	2.47
2.5	8150	4696.4	783.5	16.68	7.30
3.0	10250	5845.4	1977.0	33.82	6.10
3.5	11600	6593.2	2300.0	34.88	5.70
4.0	11600	6593.2	2700.0	40.95	2.50
4.5	11600	6593.2	2800.0	42.97	0.50
5.0	11600	6593.2	2800.0	42.47	

[a]Data from Maksymowych (1959). [b]Data from Michelini (1958)

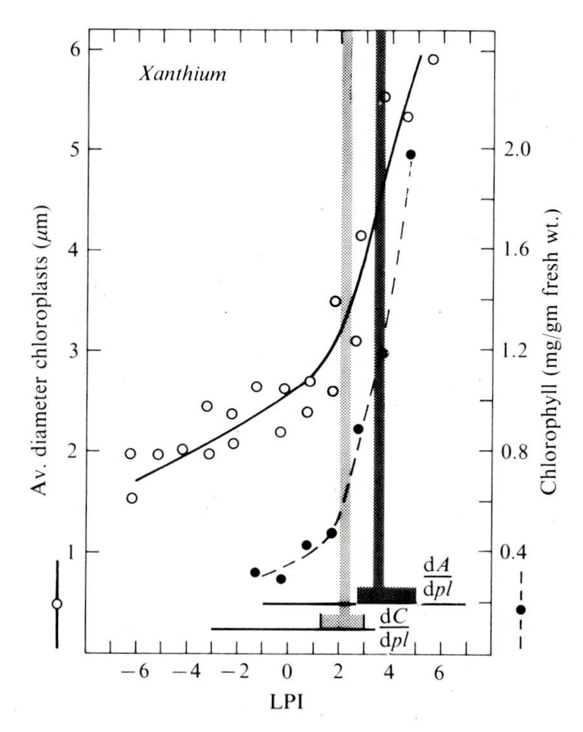

Fig. 55. Chloroplast growth in size expressed by the average diameter in μm and chlorophyll accumulation (mg/gm fresh wt.) in *Xanthium* leaves are plotted as a function of the morphological age of leaves. Shaded areas indicate time of maximum rate of cell division and growth of leaf area. A definite relationship exists between growth and cellular enlargement of the lamina and the growth in diameter of the chloroplasts within its constituent cells. (From Holowinsky, Moore and Torrey, *Protoplasma*, **60**, 1965.)

list approximately 1.2 to 1.4 μg of chlorophyll per mg fresh weight of tissue at LPI 4.0, which is the stage of near-maturity of the lamina. It can be estimated from Michelini's data that mature leaves yield approximately 0.21% chlorophyll of fresh weight and 1.4% chlorophyll of dry weight. These values compare favorably with chlorophyll (*a*, *b*) of *Pelargonium*, *Helianthus* and *Ailanthus* leaves (Rabinowitch, 1945) for dry weight, and many other genera quoted by Lubimenko (1928), for fresh weight.

Perhaps it is more helpful to follow chlorophyll synthesis on a single cell basis at various stages of cellular differentiation. To achieve this (Table 6, Fig. 56), the amount of chlorophyll in a leaf was divided by the average number of photosynthetic cells at corresponding LPI values. A meristematic cell at LPI −2.0 contains 2.18 pg of chlorophyll, whereas a mature leaf cell has about 42 pg or approximately 2.8×10^{10} chlorophyll molecules. This gives about a nine-fold increase in the amount of chlorophyll in the transition

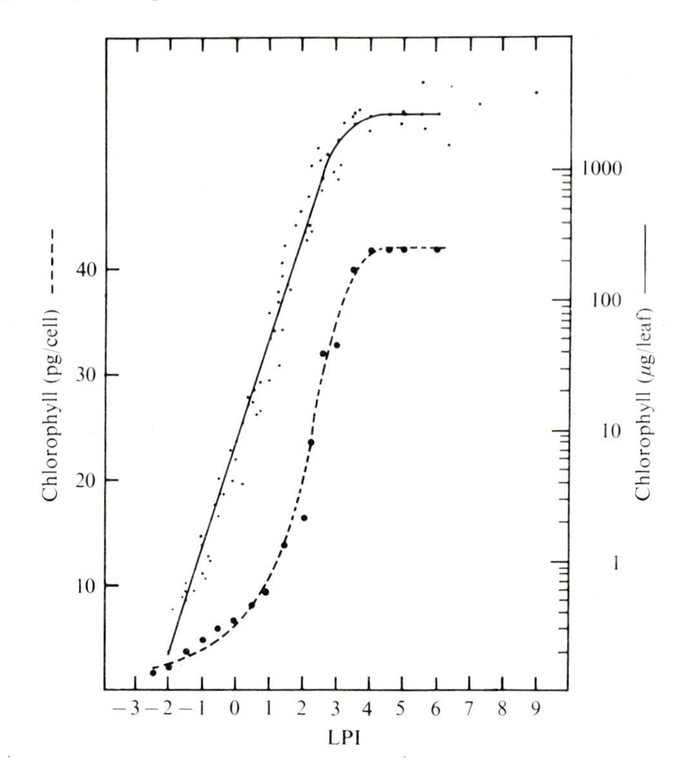

Fig. 56. The amount of chlorophyll expressed in pg per cell and the total chlorophyll (μg) per leaf on the right ordinate plotted as a function of the morphological age of leaf (LPI). A mature leaf yields approximately 0.21 % chlorophyll of fresh weight and 1.4 % chlorophyll of dry weight. A meristematic cell contains approximately 2.2 pg of chlorophyll whereas fully differentiated cell has about 42 pg or roughtly 2.8×10^{10} chlorophyll molecules. The maximum rate of chlorophyll synthesis (LPI 3.0) coincides well with the time or rapid cellular differentiation. (From Michelini, *Amer. Jour. Bot.* **45**, 1958; Maksymowych, *Amer. Jour. Bot.* **46**, 1959.)

from meristematic to mature condition. The maximum rate of chlorophyll synthesis (Table 6) on a per cell basis is between LPIs 2.5 and 3.0 which coincides well with the time of rapid cellular differentiation.

The chlorophyll content of the cell must be closely associated with photosynthetic activity because the photosynthetic rate is proportional to the chlorophyll concentration within the cell. The chlorophyll data substantiates a conclusion that a mature and most efficient photosynthetic cell apparatus is completed around LPI 6.0.

20

Respiration

Michelini (1958) studied respiration in *Xanthium* by following the oxygen consumption during the course of leaf development. As evident from Fig. 57, rates of oxygen uptake expressed on a leaf basis are low in the early development. At LPI 0.0 when the leaf is 10 mm long, it consumes about 5 μl O_2/h. During the subsequent growth, between LPIs 0.0 and 2.5, an exponential increase in respiratory rates is noticeable. This increase obviously is associated with increased rates of cell division at early plastochrons followed by cell enlargement and synthesis of more tissue during later plastochrons. An average mature leaf consumes about 580 μl of oxygen/h. It should be indicated that the logarithmic plot of respiration in Fig. 57 gives a straight line during the early development. A conventional non-logarithmic plot of O_2 uptake vs. LPI will result in a familiar sigmoid curve with a rapid increase in respiratory rates between LPIs 1.0 and 3.0.

It is customary to present the respiratory rates in terms of Q_{O_2} expressed in millilitres of oxygen uptake per gram of the fresh weight per hour. The estimated Q_{O_2} for *Xanthium* leaves are as follows: 0.65, 0.62 and 0.45 at LPIs 0.0, 2.0 and 7 respectively. This fits well within the range of other genera. Stiles and Leach (1952) estimated 0.093 Q_{O_2} for *Verbascum* and 0.8 Q_{O_2} for *Papaver*.

More meaningful information can be obtained if respiratory data are presented on a 'per cell' basis. Such data would give information concerning the respiratory activity of meristematic, differentiating and mature cells during the course of leaf development. Rates of oxygen consumption on a per cell basis are plotted in Fig. 57 vs. leaf plastochron age. These rates were estimated using the combined data of Michelini (1958) and Maksymowych (1959). Evidently, a meristematic cell has a low rate of oxygen consumption, only about 1.6 pl/h. At LPI 2.5 an enlarging cell manifests progressively higher respiratory rate. A mature leaf cell consumes about 5.5 pl O_2/h, which is about 3.5 times higher than the meristematic cell.

It would be of interest to make a quantitative comparison of leaf data with meristematic, differentiating and mature root cells. Brown and Broadbent (1950) give the following values: 11.9 pl O_2/h for a meristematic cell, 27.8 pl/h for an enlarging cell and 59.2 pl/h for a mature cell of *Pisum sativum*. Jensen (1955) lists 6.3 pl O_2/h for a meristematic cell and 32.4 pl O_2/h for a

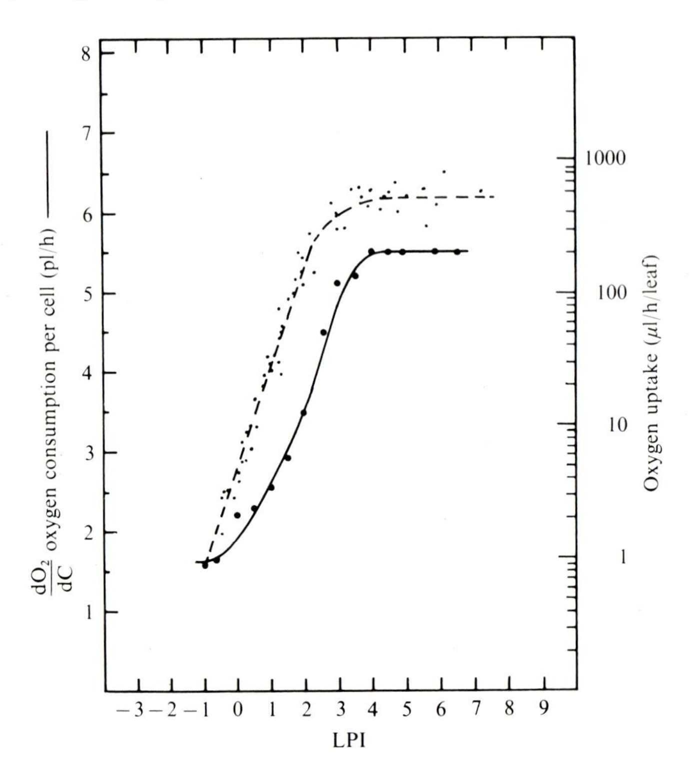

Fig. 57. Rate of oxygen consumption per cell (pl/h) and oxygen uptake per leaf (μl/h) on the right ordinate plotted vs. LPI. An average mature *Xanthium* leaf consumes about 58 μl of O_2 per hour. A meristematic cell has a low rate of O_2 consumption (1.6 pl/h); a mature leaf cell consumes about 5.5 pl O_2 per hour. (From Michelini, *Amer. Jour. Bot.* **45**, 1958; Maksymowych, *Amer. Jour. Bot.* **46**, 1959.)

mature *Vicia faba* root cell. Goddard and Bonner (1960) present some respiratory data for *Zea mays* roots. The rate of oxygen consumption in meristematic cells at 1 mm distance from the tip is about 8 pl/h. Enlarging cells have progressively higher rates and mature cells beyond 10 mm distance from the tip consume about 47 pl O_2/h. In each case the authors list approximately a six-fold respiratory increase during the transition from a meristematic to a mature condition.

By comparison the respiratory rates of leaf cells are relatively low. It can be argued, however, that the primary root is a different and a faster growing organ and is therefore more active in respiratory metabolism. Oxidative ATP synthesis in root cells may be partially responsible for higher respiratory rates. Because of photosynthetic ATP synthesis in leaf cells, the energy requirement of the two organs may be different. In addition, there must be some variability of respiratory rates associated with various types of tissues. For instance, the epidermal cells or parenchyma bundle sheath cells may have significantly different respiratory rates than spongy mesophyll cells. Unfortunately, no data is available to compare the respiratory metabolism of these tissues.

21

Correlation of DNA synthesis, cell division, cell differentiation and enzymatic activity

Leaf development has been studied (Avery, 1933; Foster, 1936; Esau, 1965; Denne, 1966) primarily from the morphological point of view, but the metabolic aspects related directly to development are experimentally more difficult to approach and are not as well understood. An attempt is made to correlate DNA synthesis and cell division with cellular differentiation and enzymatic activity, emphasizing their quantitative and temporal aspects.

As indicated, the percent of labeled nuclei is a relative measure of DNA synthesis, since ^{3}H-thymidine is incorporated into nuclei of the cells synthesizing DNA prior to mitosis. The DNA synthesis curve in Fig. 58 (Maksymowych and Kettrick, 1970) starts approximately at LPI -0.5 with about 20% of the nuclei labeled; it gradually decreases during subsequent development and stops at LPI 2.8. This means that DNA synthesis in young leaves of LPI -0.5 is high. The synthetic activity decreases gradually during the course of development and stops at LPI 2.8 when the leaf is about 70 mm long.

The assessment of mitoses between LPIs -4 and -3 was about 6%. This indicates high mitotic activity in the early stages of lamina development. Cell division gradually decreases in the course of development and stops at LPI 2.8. The curve profiles for cell division and DNA synthesis are very similar. Perhaps such correlation should be expected, since both processes are part of the cell cycle.

The rate of increase in cell volume represented in Fig. 58 on the right ordinate can be regarded as some measure of cellular differentiation. During the early part of development between LPIs -4 and -0.5, there is little enlargement and the lamina consists basically of six layers of meristematic cells, roughly 10 μm long and 8 μm wide, as can be visualized in a cross section taken from the middle portion of the lamina. This is the period of highest mitotic activity and DNA synthesis in which the leaf lamina grows primarily by cell multiplication. Between LPIs -0.5 and 2.8 the number of cells engaged in DNA synthesis and the number of cell divisions decrease steadily while cell enlargement increases. Cell enlargement has its maximum at about LPI 3.5 and is completed between LPI 5 and 6.

According to the above observations, leaf development can be divided

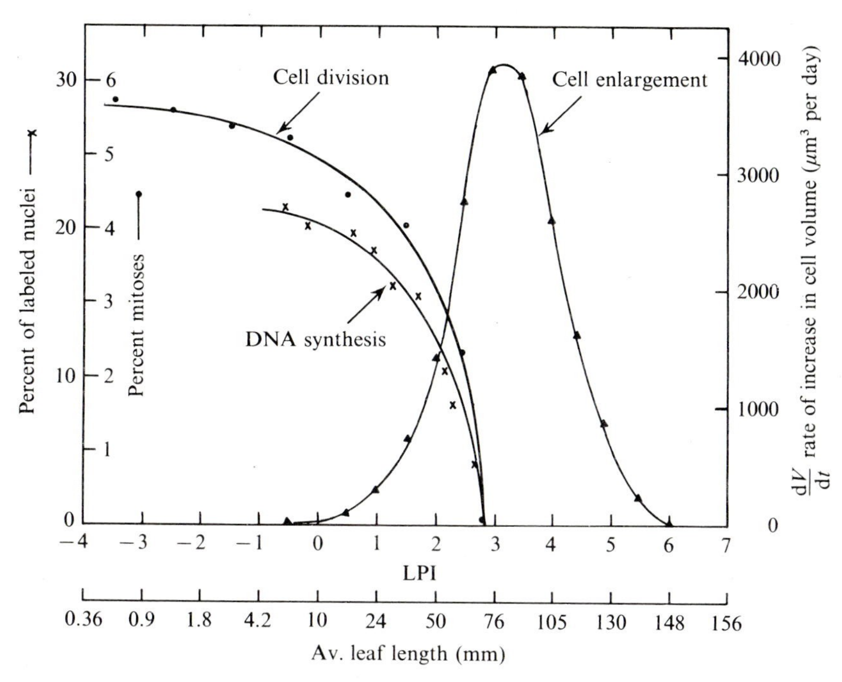

Fig. 58. Percent of labeled nuclei and percent mitosis plotted vs. leaf plastochron index. The absolute rate of increase in cell volume is plotted against LPI on the right ordinate. Average leaf length is indicated at corresponding LPI value. DNA synthesis and cell division are highest during the early stages of leaf development, they stop around LPI 3.0. Rates of cell enlargement are negligible during the early development. Maximum cell enlargement and cell differentiation is reached after cessation of cell division. Cells are fully differentiated between LPIs 5 and 6. Leaf development can be divided into two stages, the first in which cell division predominates, the second in which cell enlargement predominates. The two stages overlap between LPIs 0.5 and 2.8.

into two stages, the first in which cell division predominates and the second in which cell enlargement predominates. The first one lasts from early development up to LPI 2.8, the second from LPI −0.5 to about 6. The two stages overlap between LPIs −0.5 and 2.8.

Since no ^{3}H-thymidine was incorporated into nuclear DNA after the cessation of cell division in the leaf lamina, metabolic or endomitotic DNA synthesis after LPI 3.0 seems improbable. A different situation is found in the root system (Torrey, 1965) where progressive polyploidization and endomitoses occur along the length of a primary root in the direction of the more differentiated cells. It should be noted, however, that the exclusion of any possible endomitotic DNA synthesis in *Xanthium* leaves is based only on ^{3}H-thymidine incorporation. Cytospectrophotometric determination of nuclear DNA would provide additional helpful data. It appears also that

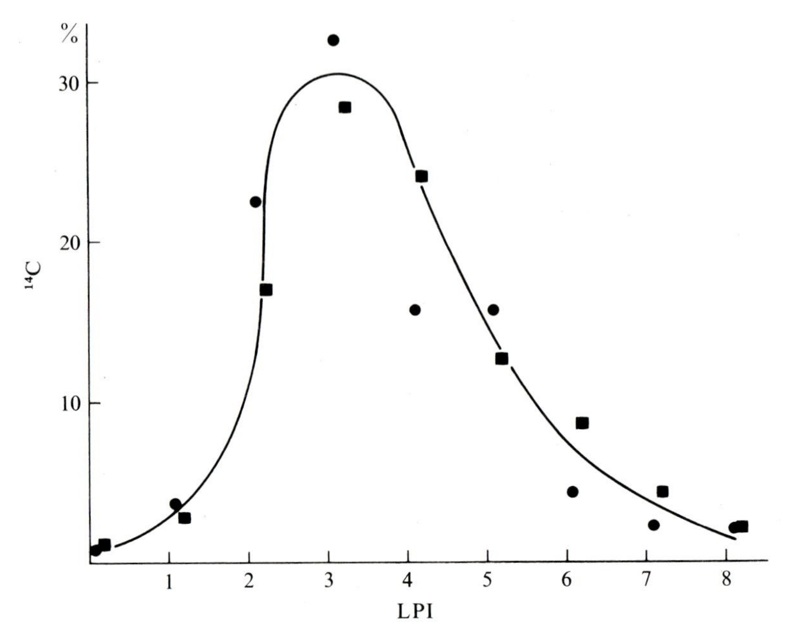

Fig. 59. Distribution of radiocarbon in the soluble proteins of the individual *Xanthium* leaves after photosynthesis in $^{14}CO_2$. The peak of protein synthesis, between LPIs 3 and 4, coincides with the maximum expansion in cell volume and rapid cellular differentiation. (From Loewenberg, *Plant and Cell Physiol.* **11**, 1970.)

no DNA synthesis is necessary for enlarging and differentiating *Xanthium* leaf cells. Evidence for such conclusion is based on cell division data, autoradiographic studies of ^{3}H-thymidine incorporation into nuclear DNA, and rates of cell enlargement of intact leaves. Studies by Kettrick (1969) on the excised parts of *Xanthium* leaves provide additional evidence. Cultivated leaf discs of LPI 0.5 to 1.0 grew exclusively by cell enlargement for about nine days. During this period, neither cell division activity nor thymidine incorporation into DNA was found. From inability of the tissue to incorporate ^{3}H-thymidine into DNA it can be concluded that DNA synthesis was not needed for cell enlargement.

Leaf development involves orderly changes in the composition of many enzymes and proteins. Proteins are of particular interest because they may be involved in enzymatic activity, thus indirectly regulating cellular differentiation and morphogenetic processes.

Loewenberg (1970) studied incorporation of $^{14}CO_2$ into the soluble proteins of developing *Xanthium* leaves. The plants were supplied 0.1% carbon dioxide with 0.4 mCi $^{14}CO_2$. After two hours of growth in an illuminated chamber more than 95% of the $^{14}CO_2$ had been taken up by the

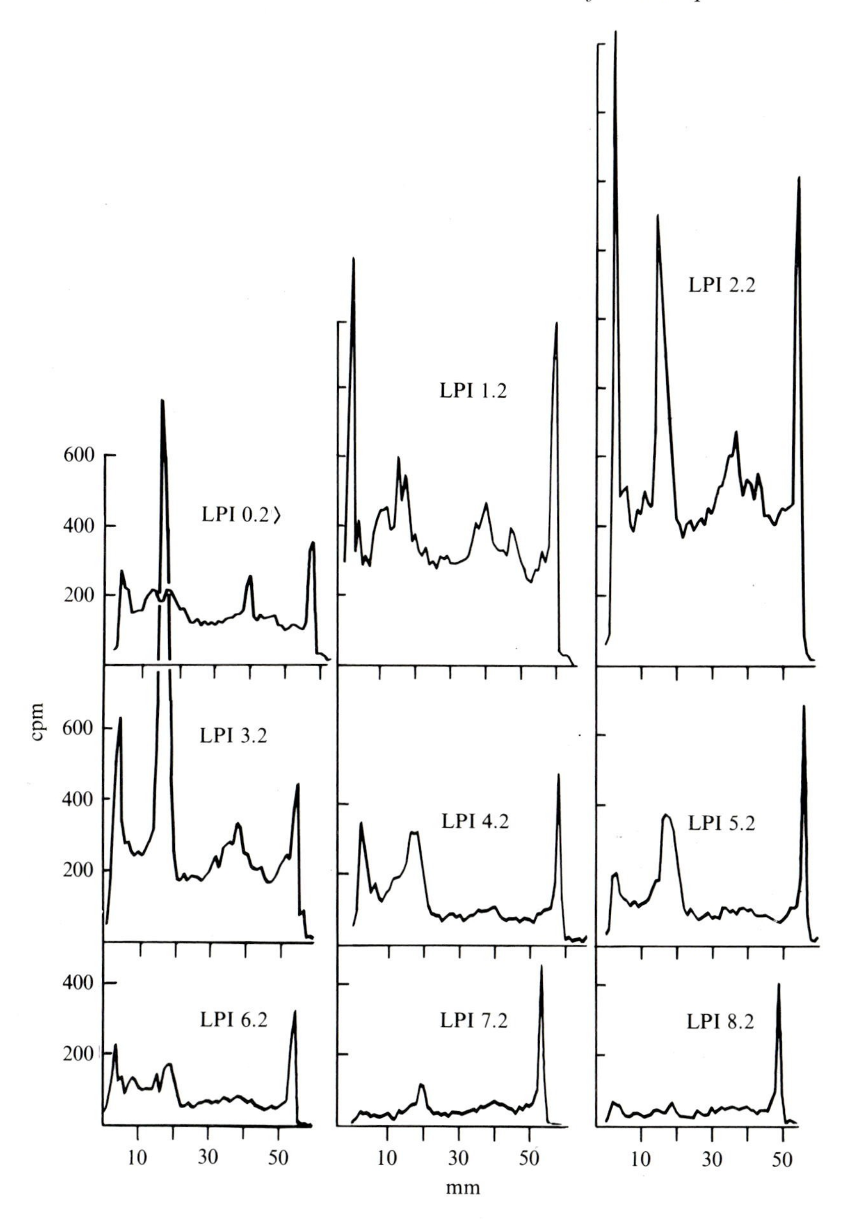

Fig. 60. Radiocarbon distribution in the gels of the soluble proteins, electrophoretically separated, of the leaves from a plant 14.2 plastochrons old after photosynthesis in $^{14}CO_2$. Origin of the cathode is at left. The maximum peak is found at LPI 3.2. (From Chen, Towill and Loewenberg, *Physiol. Plantarum,* **23**, 1970.)

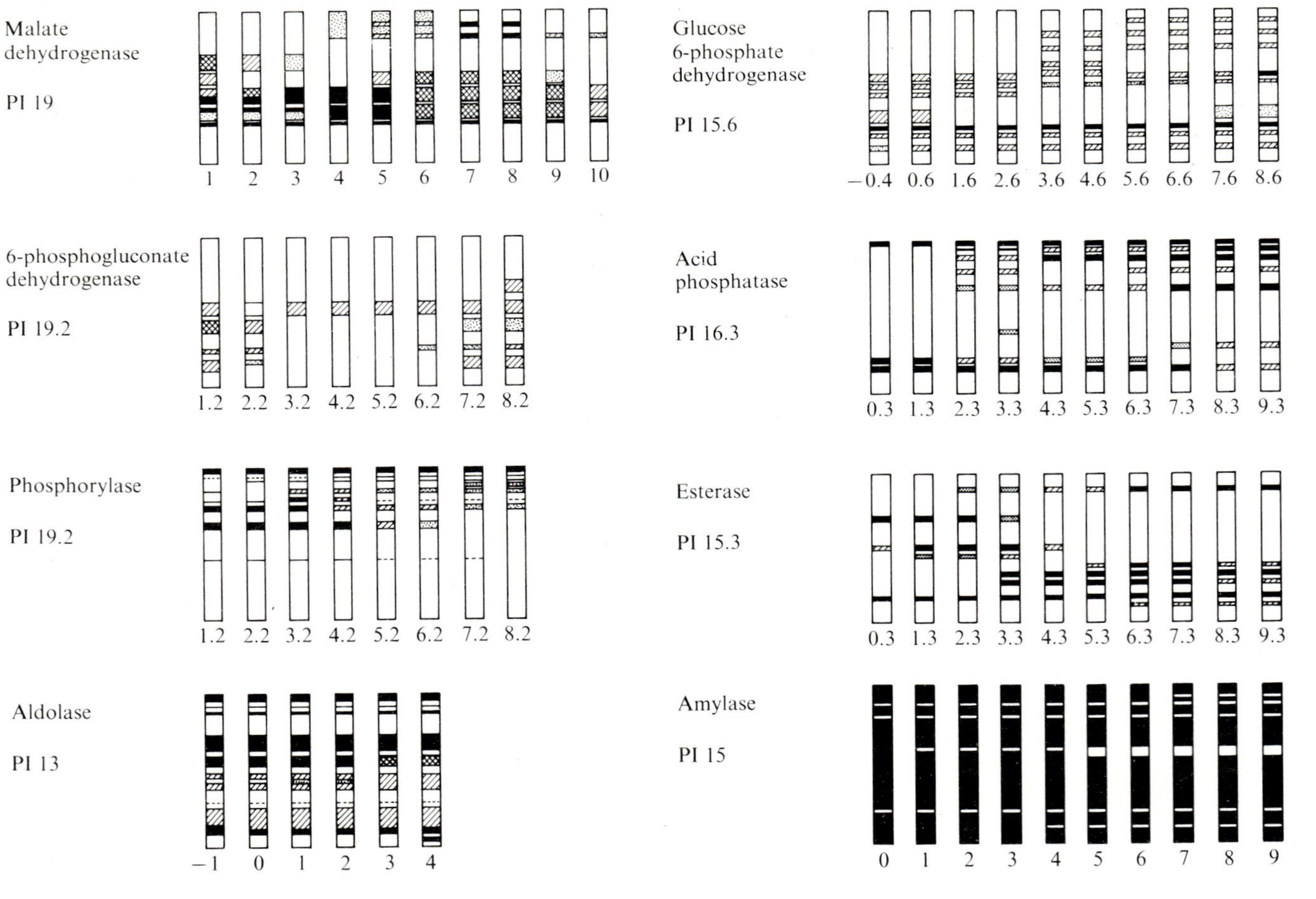

Fig. 61. Diagrammatic zymograms represented at various stages of development. Numbers at the bottom of each zymogram are LPI units. Some isoenzymes appear only in very young leaves with active cell division and DNA synthesis, others in rapidly expanding or in mature leaves. Many isoenzymes were active during the whole course of leaf development. (From Chen, Towill and Loewenberg, *Physiol. Plantarum* **22**

leaves. The soluble proteins were separated by acrylamide-disc electrophoresis and after elution the radioactivity was measured by means of liquid-scintillation spectrophotometry.

The leaves that most actively incorporated $^{14}CO_2$ into their soluble protein were about three plastochrons old (Fig. 59). This is the stage of the most rapid cellular differentiation. Younger leaves with active cell division and mature leaves incorporated significantly smaller amounts of $^{14}CO_2$.

The rate of fixation of $^{14}CO_2$ into soluble proteins correlates well with cellular differentiation. Young leaves at LPI 1.0 (Fig. 58) show little increase in cell volume and have small rates of protein synthesis. The peak of differentiation is reached around LPI 3.0 and the maximum rate of protein synthesis is also around at this stage.

The incorporation of $^{14}CO_2$ and protein synthesis in fully differentiated leaves can be attributed to chloroplast growth which is known to continue after LPI 4.0 (Holowinsky *et al.* 1965).

The distribution of the radiocarbon incorporated into the soluble leaf proteins is presented in Fig. 60. Radioactivity of the younger leaves has spread along the entire length of the acrylamide gels. The most prominent peak has emerged at LPI 3.2; its height decreased rapidly in older leaves.

Proteins associated with this peak can be related to an increased metabolism of differentiating cells and chloroplast development. Incorporation of $^{14}CO_2$ into soluble proteins decreased gradually as the leaves matured. The gradual inability to produce proteins is an indication of a progressive senescence.

Chen, Towill and Loewenberg (1970) studied the appearance of various isozymes in developing *Xanthium* leaves. They found that the isozyme composition of a leaf was determined by leaf age and plant age. Changes in the isozyme patterns coincided with the cessation of cell division or with the completion of leaf growth. Many of the isozyme patterns changed markedly at about LPI 3.0. This was evident in the glucose 6-phosphate dehydrogenase, phosphorylase, aldolase, and 6-phospho-gluconate dehydrogenase. At LPI 6.0 the leaves had completed their growth and some isozymes (Fig. 61) were most prominent in such older leaves. This was the case with acid phosphatase, esterase, amylase and guaiacol peroxidase. Changes of enzymatic activity correlated with various stages of development indicate the regulatory nature of enzyme involvement in developmental processes. For instance, one of the glucose 6-phosphate dehydrogenases was associated with very young leaves with active cell division and DNA synthesis, another with rapidly expanding leaves, and another with still older leaves. However, the number of amylase isozymes increased gradually with leaf age.

The importance of the correlation of enzymatic activity with development becomes apparent. De-novo synthesis and activation of various enzymes, along with changes in metabolism seem to be part of the regulatory mechanism of leaf development. Whether this de-novo synthesis and activation are under hormonal control still remains an experimental challenge. Further complexity becomes apparent from the fact that cell metabolism must be regulated differently on a tissue level, since DNA synthesis in the palisade layer continues longer and at significantly higher rates than in the epidermal and mesophyll cells of the lamina.

22

Plant hormones in leaf development

It is difficult to give any clear account of the roles of hormonal involvement in the growth of leaves. Relatively little experimental evidence is available pertaining specifically to various stages of leaf development. It is known that auxins, gibberellins, cytokinins, abscisic acid and other growth substances usually affect such processes as cell division, cell enlargement and differentiation. Much work on these processes was done in other organs than leaves but frequently it is found that the specific hormonal effects are similar in many organs.

From a developmental point of view, it is interesting that young expanding leaves synthesize and export to the stem considerably greater amounts of both gibberellins and auxins than mature leaves. This was demonstrated by Jones and Phillips (1966) in sunflower leaves and by Wetmore and Jacobs (1953) in *Coleus* leaves. The amount of gibberellins diffusing out of sunflower leaves is high in young apical leaves but it decreases in progressively older leaves. A young *Coleus* leaf may contain over 30×10^{-8} mg of IAA, whereas in a mature leaf this value may drop to about 5×10^{-8} mg. In addition, the auxin and gibberellin contents of leaves can be positively correlated with their growth rates. It appears that these hormones are somehow involved in the control of leaf development.

The leaf lamina is an active center of auxin and gibberellin synthesis. The stem tissues, however, depend upon leaves for a supply of these growth substances. It has been an established fact that apical dominance is maintained by auxin, whereas both of these hormones control internode elongation.

That gibberellic acid and indole-3-acetic acid are involved in regulation of growth and cellular differentiation was demonstrated by Kaufman, Petering and Adams (1969). Gibberellic acid blocked cell division activity and caused a marked increase in cell lengthening in the intercalary meristem of the *Avena* internodes. Auxin promoted lateral expansion in comparable intercalary meristem cells, especially in the vicinity of vascular bundles underlying the epidermis. It also altered the plane of cell division in differentiating stomata. At physiological concentration, auxin caused a significant reduction of gibberellin-promoted growth. Auxin can be considered to be a gibberellin antagonist in growth of the excised *Avena* stem segments. This overriding effect was demonstrated only at high IAA concentration (10^{-3} and 10^{-4}M), while at low auxin concentration, the gibberellin growth effect was clearly dominant.

Much of the experimental evidence concerning the various effects of plant hormones, and their regulatory nature on growth and development, comes from tissue culture work. Cultured leaves often need to be provided with an exogenous supply of cytokinin to satisfy their growth requirements. In contrast to auxins and gibberellins, cytokinins are not synthesized in the leaf. It appears that the root apical meristem is the primary organ involved in cytokinin synthesis. Such conclusion is based on the fact that roots are required for maintainance of a steady state system of protein synthesis and chlorophyll levels in leaves. This dependance can be eliminated by supplying the leaves with cytokinin. It has been established that cultured leaves will remain green and healthy for a long time if cytokinins are provided in the medium. Additional evidence comes from xylem sap analysis which revealed that small quantities of cytokinins are present in the ascending xylem sap. It has not been established with certainty, however, that the root apical meristem is the only source of cytokinin synthesis and supply.

Abscisic acid also known as dormin or growth inhibitor is synthesized in mature leaves. It may be involved in such processes as leaf senescence and abscission. This growth regulator accelerates senescence, leaf abscission and in some cases may completely arrest apical growth if applied to young developing shoots.

23

Senescence

After reaching maturity, the leaf remains functional for about 15 to 20 plastochrons and then it enters a period of senescence. Many physiological and biochemical changes are associated with leaf senescence (Leopold, 1961). The photosynthetic capacity, as measured by the rates of CO_2 fixation, gradually decreases. Pronounced changes in carbohydrate and protein metabolisms are evident and the auxin level is significantly reduced. Finally, decomposition of the chlorophyll pigments and yellowing of the blade become apparent. As all these metabolic changes take place, there is a translocation of many organic and inorganic molecules from the leaf to the juvenile shoot, until abscission totally interrupts this export.

While the mechanism controlling senescence is not well understood, there is a large body of evidence that auxin, kinetin, nucleic acids and proteins are intimately involved in this process. Experiments with excised leaves have shown that the decrease in protein content of the blade is not necessarily due to an inability of the leaf cells to synthesize amino acids but is due, rather, to a failing ability to incorporate these amino acids into proteins.

In 1957 Richmond and Lang demonstrated that there was a considerable delay in chlorophyll degradation and protein loss if excised *Xanthium* leaves were treated with kinetin. Osborne (1962) also studied the effect of kinetin on protein and nucleic acid metabolism in *Xanthium* leaves during senescence. As can be seen in Fig. 62, the levels of chlorophyll, protein and RNA degradation were significantly reduced in leaves treated with kinetin. The DNA level was not significantly affected. It is evident from the experimental results that kinetin in some way retards the process of senescence. Osborne suggested that the impairment of RNA synthesis might be a primary feature of cellular senescence and a direct cause of protein loss. It appears that the primary action of kinetin in controlling leaf senescence involves the regulation of RNA synthesis with a resultant regulation of protein synthesis.

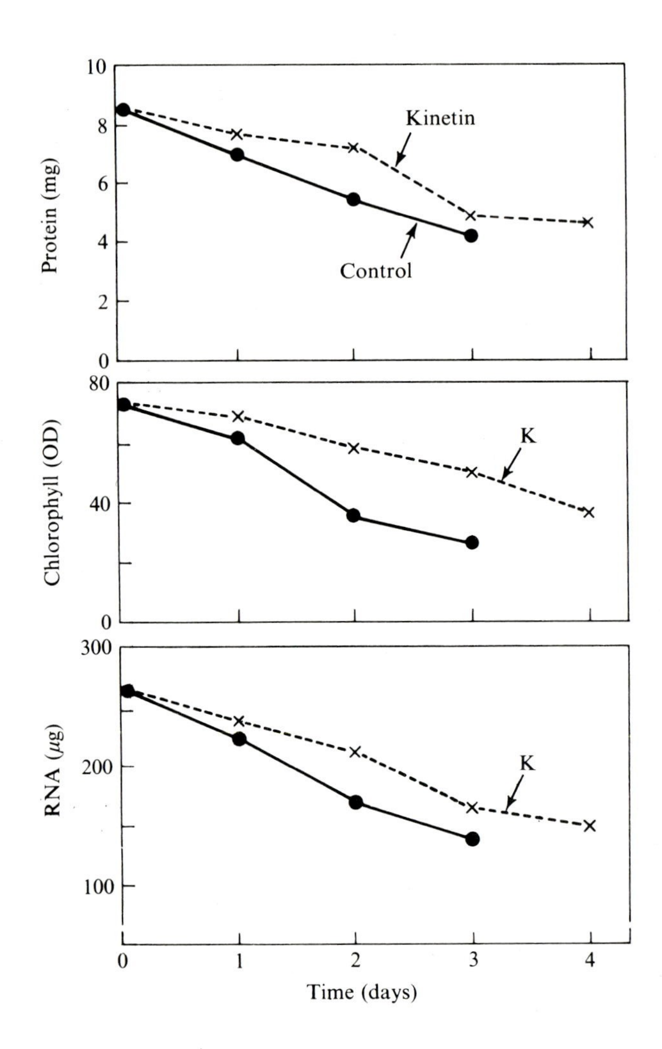

Fig. 62. The amount of RNA, protein and chlorophyll of *Xanthium* leaf discs plotted vs. time of treatment with kinetin during a period of senescence. The levels of protein, RNA and chlorophyll degradation were significantly reduced in leaves treated with kinetin. (From Osborne, *Plant Physiol.* **37**, 1962.)

24

Concluding statements

In light of the discussion presented in various chapters it becomes clearly evident that leaf development is a series of integrated processes which are controlled and regulated during the course of development. This becomes especially apparent from correlation of metabolic processes with cellular and morphological changes associated with development. The nature of the control mechanisms is obscure and little understood. An attempt will be made, however, to discuss several processes which appear to be controlled and regulated during leaf development.

It was stated that rates of cell division are highest in the early plastochrons. These rates decline gradually during subsequent development and stop at LPI 3.0. During the cell division phase only a negligible cell enlargement is noticeable and rapid enlargement appears to be suppressed until LPI 2.0, or shortly before complete cessation of mitotic division. Apparently some control mechanisms must be involved in maintaining a uniform condition of compact undifferentiated cell layers throughout the extent of the lamina from the midrib to the margin. Cellular metabolism is programmed for multiplication of cells, i.e. DNA synthesis and synthesis of the necessary precursors associated with cell duplication, and suppression of cell enlargement. Such a pattern of development is altered when small discs of meristematic lamina are cultured in basal White's medium without addition of any growth substances. In cultured condition DNA synthesis and cell division stop abruptly. A rapid cell enlargement is initiated within a day. It appears that some factor, or set of factors, becomes unavailable to cultured meristematic cells. These factors could be growth regulators supplied by the vascular system to the juvenile leaves. Cultured cells apparently are not synthesizing such growth regulators at all, or perhaps not in a sufficient concentration to maintain prolonged mitotic activity. Cytokinins are known to enhance cell division in tissue culture. It is not unreasonable to assume that they are the growth regulators active during the cell division phase in developing *Xanthium* leaves.

Cellular metabolism is strongly correlated with developmental stages of the leaf. A meristematic cell maintains a low respiratory rate, about one-fourth of the mature cell. Photosynthetic rates are also negligible since chloroplasts and stomata are not fully developed. It is logical to assume that the vascular tissue must supply the meristematic cells with many precursors and respiratory substrates. In contrast, however, the nucleic acid metabolism of meristematic cells is high. This is manifested by high rates of incorporation of ^{3}H-thymidine into nuclear DNA and high rates of cell division. In other words, the metabolism of young leaves is programmed for cell multiplication. Whether growth substances like cytokinins are directly responsible for maintaining a high nucleic acid metabolism in leaf cells remains an experimental challenge.

It was established that DNA synthesis in the palisade layer continues for at least 3.5 days longer and at a significantly higher rate than in the epidermal and mesophyll cells of the lamina. Considering also cellular growth on a comparative basis, the palisade cells start elongating in height at least one plastochron sooner and at a much higher rate than the epidermis. The epidermal cells enlarge at higher absolute and relative rates than palisade mesophyll. From the above considerations of differential rates, the inference can be made that cell metabolism is regulated on a tissue level and that cell enlargement and differentiation are programmed also on a tissue level.

After LPI 2.0, a shift in lamina development was demonstrated. DNA synthesis and cell division decline gradually and stop around LPI 3.0. A rapid cell enlargement, possibly under auxin control, takes place. DNA synthesis stops and cellular metabolism is shifted from cell reproduction to cell differentiation. New cell wall materials, primarily hemicellulose and lignins, are synthesized to satisfy the needs of the rapidly increasing cell volume. Even exogenously supplied tritiated thymidine is metabolized and its labeled fraction is incorporated into the secondary wall of differentiating cells. This example indicates the ability of a cell to regulate and adjust its metabolism to existing needs.

The rapid chloroplast growth coincides with the maximum rate of increase in leaf area and the maximum rate of cell enlargement. Again, it appears that the chloroplast development and its function are closely integrated with cellular and organ development. It is reasonable to assume that nuclear control is involved in chloroplast development.

After LPI 4.0, the leaf lamina becomes mature both anatomically and physiologically. Stomata are fully developed and the intercellular spaces are established in the spongy mesophyll to facilitate CO_2 diffusion. A mature photosynthetic organ has been established.

There is ample evidence that there are growth substances that both enhance and suppress the division and enlargement of cells. However, one cannot yet specify the mode of action of these control substances, i.e. how they regulate the expression of cellular genetic constitution and translate this information for a programmed development.

Literature cited

Avery, G. S., Jr. (1933). Structure and development of the tobacco leaf. *Amer. Jour. Bot.* **20**, 565–92.

Bernier, G. (1966). The morphogenetic role of the apical meristem in higher plants. *Les congrés et colloques de L'université de Liège*, **38**, 151–211.

Brown, R. and D. Broadbent. (1950). The development of cells in the growing zones of the root. *Jour. Exp. Bot.* **1**, 249–63.

Buvat, R. (1952). Structure, évolution et fonctionnement du méristème apical de quelques Dicotyledones. *Amer. Sci. Nat. Bot.* XI, **13**, 199–300.

Buvat, R. (1955). Le méristème apical de la tige. *Ann. Biol.* **31**, 595–656.

Chen, S. L., L. R. Towill, and J. R. Loewenberg. (1970). Isoenzyme patterns in developing *Xanthium* leaves. *Physiol. Plantarum*, **23**, 434–43.

Clowes, F. A. L. (1961). *Apical meristems*. Oxford: Blackwell.

Cutter, G. E. (1964). Observation on leaf and bud formation in *Hydrocharis*. *Amer. Jour. Bot.* **51**, 318–24.

Denne, P. M. (1966). Leaf development in *Trifolium repens*. *Bot. Gaz.* **127**, 202–10.

Dormer, P., W. Brinkmann, A. Stieber and W. Stich. (1966). Automatische Silberkornzahlung in der Einzelzell-Autoradiographie. Eine neue, photometrische Methode für die quantitative Autoradiographie. *Klin Wschr.* **44**, 477.

Erickson, R. O. (1960). Nomogram for the plastochron index. *Amer. Jour. Bot.* **47**, 350–1.

Erickson, R. O. (1966). Relative elemental rates and anisotropy of growth in area: a computer programme, *Jour. Exp. Bot.* **17**, 390–403.

Erickson, R. O. and F. J. Michelini. (1957). The plastochron index. *Amer. Jour. Bot.* **44**, 297–305.

Esau, K. (1965). *Plant anatomy*, 2nd ed. New York: Wiley.

Foster, A. S. (1936). Leaf differentiation in angiosperms. *Bot. Rev.* **2**, 349–72.

Fuchs, C. M. (1966). Observations sur l'extension en largeur du limbe foliaire du *Lupinus albus*. *L.D.R. Acad. Sc. Paris* Sér. D, **263**, 1212–15.

Gifford, E. M., Jr. (1951). Early ontogeny of the foliage leaf in *Drimys winterei* var. *chilensis*. *Amer. Jour. Bot.* **38**, 93–105.

Gifford, E. M., Jr. (1963). Developmental studies of vegetative and floral meristems. *Meristems and Differentiation; Brookhaven Symposia in Biology*, no. 16, 126–35.

Girolami. G. (1954). Leaf histogenesis in *Linum usitatissimum*. *Amer. Jour. Bot.* **41**, 264–73.

Goddard, D. R. and W. D. Bonner. (1960). *Cellular respiration in plant physiology*, vol. 1. F. C. Steward, ed. New York: Academic Press. Pp. 209–312.

Goodwin, R. H. and C. J. Avers. (1956). Studies on Roots III. An analysis of root growth in *Phleum pratense* using photomicrographic records. *Amer. Jour. Bot.* **43**, 479–87.

Hearney, E. Y. (1965). Autoradiographic studies of the mitotic cycle in *Xanthium pennsylvanicum*. M.S. Thesis, Villanova University.

Hejnowicz, Z. (1955). Growth distribution and cell arrangement in apical meristems. *Acta Soc. Bot. Poloniae*, **24**, 583–608.

Holowinsky, A. W., P. B. Moore and J. G. Torrey. (1965). Regulatory aspects of chloroplast growth in leaves of *Xanthium pennsylvanicum* and etiolated red kidney bean seedling leaves. *Protoplasma*, **60**, 94–110.

Jacobs, W. P. and I. B. Morrow. (1957). A quantitative study of xylem development in the vegetative shoot apex of *Coleus*. *Amer. Jour. Bot.* **44**, 823–42.

Jensen, W. A. (1955). A morphological and biochemical analysis of the early phases of cellular growth in the root tip of *Vicia faba*. *Exp. Cell Res.* **8**, 506–22.

Jensen, W. A. (1962). *Botanical Histochemistry*. San Francisco: W. H. Freeman.

Jones, R. L. and I. D. J. Phillips. (1966). Organs of gibberellin synthesis in light grown sunflower plants. *Plant Physiol.* **41**, 1381–6.

Juniper, B. E. (1959). The surface of plants. *Endeavour*, **18**, 20–5.

Kaplan, D. R. (1970). Comparative foliar histogenesis in *Acorus calamus* and its bearing on the phyllode theory of monocotyledonous leaves. *Amer. Jour. Bot.* **57**, 331–61.

Kaufman, P. B., L. B. Petering and P. A. Adams. (1969). Regulation of growth and cellular differentiation in developing *Avena* internodes by gibberellic acid and indole-3-acetic acid. *Amer. Jour. Bot.* **56**, 318–27.

Kettrick, M. A. (1969). Autoradiographic studies of DNA biosynthesis in cultured leaf tissue of *Xanthium pennsylvanicum*. M.S. Thesis, Villanova University.

Khudairi, A. K. and K. C. Hamner. (1954). The relative sensitivity of *Xanthium* leaves of different ages to photoperiodic induction. *Plant Physiol.* **29**, 251–7.

Lance, A. (1952). Sur le structure et le fonctionnement du point végétatif de *Vicia faba*. *Amer. Sci. Nat. Bot.* XI, **13**, 301–39.

Leopold, A. C. (1961). Senescence in plant development. *Science*, **134**, 1727–32.

Loewenberg, J. R. (1970). Protein synthesis in *Xanthium* leaf development. *Plant and Cell Physiol.* **11**, 361–5.

Lubimenko, V. N. (1928). *Rev. gen. Botan.* **40**, 88, 146, 415; quoted by Rabinowitch (1945).

Maksymowych, R. (1959). Quantitative analysis of leaf development in *Xanthium pennsylvanicum*. *Amer. Jour. Bot.* **46**, 635–44.

Maksymowych, R. (1962). An analysis of leaf elongation in *Xanthium pennsylvanicum* presented in relative elemental rates. *Amer. Jour. Bot.* **49**, 7–13.

Maksymowych, R. (1963). Cell division and cell elongation in leaf development of *Xanthium pennsylvanicum*. *Amer. Jour. Bot.* **50**, 891–901.

Maksymowych, R. and M. K. Blum. (1966*a*). Incorporation of ^{3}H-thymidine in leaf nuclei of *Xanthium pennsylvanicum*. *Amer. Jour. Bot.* **53**, 134–42.

Maksymowych, R. and M. K. Blum. (1966*b*). Autoradiographic studies of the synthesis of nuclear DNA in various tissues during leaf development of *Xanthium pennsylvanicum*. *Developmental Biol.* **14**, 250–65.

Maksymowych, R., R. G. Devlin, M. K. Blum and Z. S. Wochok. (1967). ^{3}H-thymidine incorporation into nuclear DNA of leaf cells. *Plant Physiol.* **42**, 814–18.

Maksymowych, R. and R. O. Erickson. (1960). Development of the lamina in *Xanthium italicum* represented by the plastochron index. *Amer. Jour. Bot.* **47**, 451–9.

Maksymowych, R. and M. A. Kettrick. (1970). DNA synthesis, cell division, and cell differentiation during leaf development of *Xanthium pennsylvanicum*. *Amer. Jour. Bot.* **57**, 844–9.

Maksymowych, R. and Z. S. Wochok. (1969). Activity of marginal and plate meristems during leaf development of *Xanthium pennsylvanicum*. *Amer. Jour. Bot.* **56**, (1) 26–30.

MacDaniels, L. H. and F. F. Cowart. (1944). The development and structure of the apple leaf, *N.Y. (Cornell U.) Agric. Expt. Sta. Mem.* **258**, 1–29.

McGahan, N. W. (1955). Vascular differentiation in the vegetative shoot of *Xanthium chinense. Amer. Jour. Bot.* **42**, 132–40.

Michelini, F. J. (1958). The plastochron index in developmental studies of *Xanthium italicum* Moretti. *Amer. Jour. Bot.* **45**, 525–33.

Millington, W. F. and E. L. Fisk. (1956). Shoot development in *Xanthium pennsylvanicum*. The vegetative plant. *Amer. Jour. Bot.* **43**, 655–65.

Noack, K. L. (1922). Entwickelungsmechanische Studien an panaschierten Pelargonien. *Jahrb. Wiss. Bot.* **61**, 459–543.

Ogur, M. and Rosen, G. (1950). The nucleic acids in plant tissues. The extraction and estimation of deoxypentose nucleic acid and pentose nucleic acid. *Arch. Biochem.* **25**, 262–76.

Osborne, D. J. (1962). Effect of kinetin on protein and nucleic acid metabolism in *Xanthium* leaves during senescence. *Plant Physiol.* **37**, 595–602.

Rabinowitch, E. I. (1945). *Photosynthesis*, vol. I. New York: Interstate Publ. Pp. 409.

Richmond, A. E. and A. Lang. (1957). Effect of kinetin on protein content and survival of detached *Xanthium* leaves. *Science,* **125**, 650–51.

Rogers, A. W. (1967). *Techniques of autoradiography*. Elsevier.

Salisbury, F. B. (1955). The dual role of auxin in flowering. *Plant Physiol.* **30**, 327–34.

Salisbury, F. B. (1961). *Ann. Rev. of Plant Physiol.* Calif. U.S.A.

Schmidt, G. and S. J. Thannhauser. (1945). A method for determination of deoxyribonucleic acid, ribonucleic acid, and phosphoproteins in animal tissues. *Jour. Biol. Chem.* **161**, 83–9.

Schüepp, O. (1966). *Meristeme*. Basel–Stuttgart: Birkhauser Verlag.

Shushan, S. and M. A. Johnson. (1955). The shoot apex and leaf of *Dianthus caryophyllus. L. Bull. Torrey Bot. Club*, **82**, 266–83.

Smillie, R. M. and G. Krotkow. (1960). The estimation of nucleic acids in some algae and higher plants. *Amer. Jour. Bot.* **38**, 31–49.

Stiles, W. and W. Leach. (1952). *Respiration in Plants*, 3rd ed. New York: Wiley.

Sunderland, N. (1960). Cell division and expansion in the growth of the leaf. *Jour. Exp. Bot.* **11**, 68–80.

Torrey, J. G. (1965). Physiological basis of organization and development in the root. *Encyclopedia of plant physiology*. W. Ruhland, ed. Berlin: Springer Verlag. Pp. 1301–7.

Von Papen, R. (1935). Beitrage zur Kenntnis des Wachstums der Blattspreite. *Bot. Arch.* **37**, 159–206.

Watson, D. P. (1948). An anatomical study of the modification of bean leaves as a result of treatment with 2-4-D. *Amer. Jour. Bot.* **35**, 543–55.

Weidt, E. (1935). Die Entwicklung der Blätter der Melastomataceen *Heterotrichum macrodon* Planck, und *Clidemia hirta* Don. *Beitr. Biol. Pflanzen*, **23**, 252–81.

Wetmore, R. H., E. M. Gifford, Jr. and M. C. Green. (1959). Photoperiodism and related phenomena in plants and animals. *Amer. Assoc. Adv. Sci. Wash. D.C.* Pp. 255–73.

Wetmore, R. H. and Jacobs, W. P. (1953). Studies on abscission: The inhibiting effect of auxin. *Amer. Jour. Bot.* **40**, 272–6.

Index